Waves

Second Edition

F R Connor

PhD, MSc, BSc (Eng) Hons, ACGI,
C Eng, MIEE, MIERE, M Inst P

Edward Arnold

First published in Great Britain 1972 by
Edward Arnold (Publishers) Ltd, 41 Bedford Square, London WC1B 3DQ
Edward Arnold, 3 East Read Street, Baltimore, Maryland 21202, U.S.A.
Edward Arnold (Australia) Pty Ltd, 80 Waverley Road, Caulfield East,
Victoria 3145, Australia

Reprinted 1975, 1978, 1980, 1982, 1983, 1985
Second edition 1986

British Library Cataloguing in Publication Data

Connor, F. R.
Waves.—2nd ed.
1. Electromagnetic waves—Transmission
I. Title II. Connor, F. R. Wave transmission
530.1′4 QC665.T7

ISBN 0-7131-3567-0

Text set in 10/11 pt Times
by Macmillan India Ltd, Bangalore 25.
Printed and bound in Great Britain by
Billings and Sons Ltd, Worcester

Preface

This new edition has been renamed *Waves* for the sake of brevity, and various parts of the text have been revised or extended. Chapter 1 now includes more details on stripline and microstrip and a brief introduction to optical fibres. A new section on reflected and refracted waves in Chapter 3 is intended to serve as an introduction to optical phenomena, while Chapter 4 has been extended to cover higher-order modes and waveguide attenuation. Chapter 6 is devoted entirely to the new subject of optical communications in which important developments have taken place in optical fibres, sources and detectors during recent years. As in the earlier edition, the main text has been extended by the use of a set of Appendices to provide further details on such topics as stripline and microstrip, ferrites and fibre mode theory. Furthermore, worked examples are included to illustrate the text, together with a greater number of problems and answers. A large set of useful references is also provided for the interested reader.

The aim of the book is the same as in the first edition, though it should be pointed out that Higher National Certificates and Higher National Diplomas are now awarded by the Business and Technician Education Council. Furthermore, the Council of Engineering Institutions Examination is now the responsibility of the Engineering Council and is called the Engineering Council Examination.

In conclusion, the author would like to express his gratitude to those of his readers who so kindly sent in comments and corrections for the first edition.

1986 FRC

Preface to the first edition

This is an introductory book on the important topic of *Wave Transmission*. Electromagnetic waves play an essential part in many communication systems and the book endeavours to present basic ideas concerning the transmission of such waves in a concise and coherent manner. Moreover, to assist in the assimilation of these basic ideas, many worked examples from past examination papers are provided to illustrate clearly the application of the fundamental theory.

The first part of the book deals with the various types of transmission lines used for different applications. A general analysis follows, based on circuit ideas of voltages and currents. Subsequent chapters then consider the alternative concept of fields, which is essential for the proper understanding of waveguide transmission, and the book ends with a useful treatment of microwave theory and techniques.

This book will be found useful by students preparing for London University examinations, degrees of the Council of National Academic Awards, examinations of the Council of Engineering Institutions and for other qualifications such as Higher National Certificates, Higher National Diploma and certain examinations of the City and Guilds of London Institute. It will also be useful to practising engineers in industry who require a ready source of basic knowledge to help them in their applied work.

1972

FRC

Acknowledgements

The author sincerely wishes to thank the Senate of the University of London and the Council of Engineering Institutions for permission to include questions from past examination papers. The solutions and answers provided are his own and he accepts full responsibility for them.

Sincere thanks are also due to the publishers H. W. Peel & Co. Ltd for permission to reproduce their impedance chart shown in Fig. 2.14 and also to the City and Guilds of London Institute for permission to include questions from past examination papers. The Institute is in no way responsible for the solutions and answers provided.

Finally, the author wishes to thank the publishers for various useful suggestions and will be grateful to his readers for drawing his attention to any errors in the work.

Abbreviations

C.E.I.	Council of Engineering Institutions examinations, Part 2
C.G.L.I.	City and Guilds of London Institute examinations
U.L.	University of London, BSc(Eng) examination in Telecommunication, Part 3

Contents

Preface iii

1 Introduction 1

1.1 Two-wire open line 1
1.2 Coaxial cable 2
1.3 Stripline and microstrip 4
1.4 Waveguides 8
1.5 Optical fibres 10

2 Line transmission 12

2.1 The general line 12
2.2 Secondary line constants 14
2.3 Infinite line 15
2.4 Hyperbolic solutions 16
2.5 Practical line 16
2.6 General termination 17
2.7 Special cases 18
2.8 Line classification 20
2.9 Phase and group delay 24
2.10 Reflections on lines 26
2.11 Reflection coefficient 27
2.12 Voltage standing wave ratio (VSWR) 29
2.13 The Smith chart 30
2.14 Typical examples 33

3 Electromagnetic waves 37

3.1 Electromagnetic fields 37
3.2 Electromagnetic field theory 38
3.3 Boundary conditions 43
3.4 Reflected and refracted waves 46

4 Waveguide theory 50

4.1 Waveguide transmission 50
4.2 Phase and group velocities 52
4.3 Waveguide equation 53

4.4 Rectangular waveguides 55
4.5 Rectangular modes 55
4.6 Circular waveguides 60
4.7 Circular modes 60
4.8 Higher-order modes 61
4.9 Waveguide attenuation 63
4.10 Launching in waveguides 67

5 Microwave techniques 69

5.1 Microwave sources 69
5.2 Microwave components 73
5.3 Microwave measurements 78

6 Optical communications 88

6.1 Sources 88
6.2 Fibre cables 91
6.3 Detectors and receivers 99
6.4 Further developments 104

Problems 107

Answers 112

References 113

Appendices 115

A Stripline and microstrip 115
B Ferrites 116
C Cavity resonators 119
D Fibre mode theory 123

Index 126

Symbols

a	core radius
	waveguide width
b	core radius
	normalised propagation constant
	waveguide height
c	velocity of light
f	frequency
h	height of dielectric
	Planck's constant
i	current
	angle of incidence
$\boldsymbol{i}$	unit vector
$\boldsymbol{j}$	unit vector
k	wave number
$\boldsymbol{k}$	unit vector
k_c	the quantity $2\pi/\lambda_c$
l	length
m	any number
n	any number
	refractive index
p	any number
q	any number
r	angle of refraction
	normalised resistance
	radius
s	standing wave ratio
t	thickness
	time
v	velocity
	voltage
v_g	group velocity
v_p	phase velocity
w	strip width
x	normalised reactance
y	normalised admittance
z	normalised impedance
A	any constant
B	any constant
	bandwidth
	bit rate
C	capacitance
$\boldsymbol{D}$	electric flux density
$\boldsymbol{E}$	electric field strength
G	conductance
$\boldsymbol{H}$	magnetic field strength

I	current
$\mathbf{J}$	current density
J_m	Bessel function of order m
J'_m	first derivative of J_m
K	any constant
L	inductance
	length
P	power
R	resistance
R_s	resistivity
V	voltage
W	energy
X	reactance
Y	admittance
Z	impedance
Z_0	characteristic impedance
α	attenuation coefficient
β	phase-change coefficient
γ	gyromagnetic ratio
	propagation coefficient
δ	loss angle
	skin depth
ε	permittivity
ε_0	permittivity of free space
ε_r	relative permittivity
λ	wavelength
λ_c	cut-off wavelength
λ_g	waveguide wavelength
λ_0	free space wavelength
μ	permeability
μ_0	permeability of free space
ν	frequency
ρ	charge density
	reflection coefficient
σ	conductivity
τ_d	time-delay
ω	angular frequency
ω_0	angular resonant frequency
Δ	relative refractive index
∇	operator del

1
Introduction

The transmission of energy between a source and some distant point requires the use of a transmission medium generally called a transmission line. In certain cases this requires a physical structure, but in some cases, no structure is required as transmission is achieved directly through free space as an electromagnetic wave. Over the years, several types of transmission lines have been used, each having its own particular applications and limitations, yet finding wide use in various communication systems.

The development of transmission lines springs largely from the use of the familiar two-wire electrical power line for carrying large quantities of power from a generator to its load. However, the demands of communication systems with their far greater frequency requirements, led to the development of various other types of transmission lines which will be treated in this book and the more familiar power line will be referred to only briefly in the text.

Broadly speaking, transmission lines are either lumped lines or distributed lines. Lumped lines are so-called because their electrical parameters such as resistance, inductance and capacitance are lumped at intervals along the line and are unlike the distributed line in which these parameters are uniformly spread over the whole length of line. Most practical lines are of the uniform type as they are easily manufactured and have better characteristics than the lumped line. The most commonly used lines are the two-wire open line, coaxial line, stripline, microstrip line, waveguide and optical fibre.

1.1 Two-wire open line[1, 2]

This consists essentially of two conductors spaced a certain distance apart and used extensively in power systems, telegraphy and telephone systems and in certain areas of radio transmission. Its main use is in the lower frequency range of work and it is simple to manufacture. Such lines are chiefly characterised as having resistance per unit length, while the inductance and capacitance per unit length are usually quite small.

The commonest two-wire line is the overhead power line operating at high voltages. The distance between the conductors is large compared to the conductor diameter, but at high frequencies, radiation losses are minimised by considerably reducing the distance between conductors. Two-wire lines used for communication purposes such as antenna feeders or down-leads to receivers, have characteristic impedances between 70 Ω and 600 Ω. They are spaced apart by dielectric spacers or moulded in some dielectric material. Propagation is essentially as a TEM (transverse electric and magnetic) wave, in which the energy is carried by the fields and the wave is guided along by the conductors as shown in Fig. 1.1.

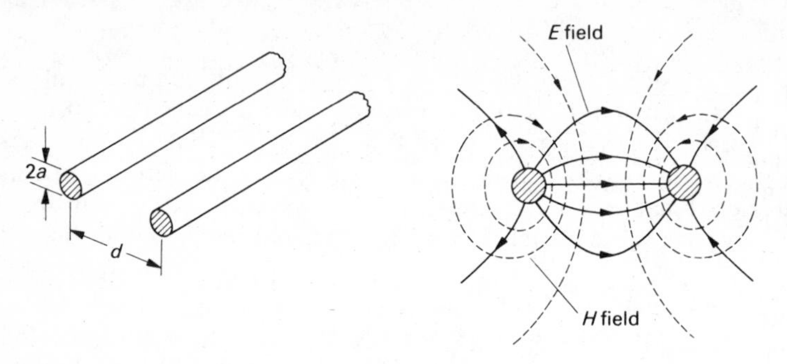

Fig. 1.1

It can be shown that the relevant parameters, R, L, C per loop metre are given by

$$R_{\text{d.c.}} = \frac{2\rho}{\pi a^2}\ \text{ohm} \qquad L = \frac{\mu_0}{\pi}\ln d/a\ \text{henry}$$

$$R_{\text{a.c.}} = \frac{1}{\pi a}\left[\frac{\omega\mu\rho}{2}\right]^{1/2}\ \text{ohm} \qquad C = \frac{\pi\varepsilon_0}{\ln d/a}\ \text{farad}$$

where ρ is the specific resistance, a is the radius of the conductors and d is the distance between their centres. Also

$$Z_0 \simeq \sqrt{\frac{L}{C}} = 276\log_{10} d/a\ \text{ohm}$$

$$v \simeq \frac{1}{\sqrt{\mu_0\varepsilon_0}} = 3\times10^8\ \text{m/s}$$

1.2 Coaxial cable[3, 4]

The two-wire line is useful mainly at the lower frequencies and up to about 100 MHz in short lengths only. At higher frequencies, serious losses occur due to skin effect in the conductors and radiation from the surface.

Due to the severe radiation losses, a closed field configuration must be used in which an inner conductor is surrounded by an outer cylindrical sheath and is known as a coaxial cable. It has the advantage that the fields are confined within the outer conductor thus eliminating radiation losses and it is also shielded from outside interference. The medium between the conductors may be either air or a dielectric material. Such coaxial cables find extensive use not only at power frequencies where the main problem is one of insulation, but also at very high frequencies for radio and television applications.

A typical air-cored cable has an inner conductor of copper, held in position by polythene discs, which is surrounded by one or more layers of steel tape for screening or strength. Flexible forms of cable are also used with a polythene dielectric and outer copper braiding for flexibility.

In the field of communications, primary considerations are those of attenuation and distortion at the frequencies of operation, while a secondary consideration is that of power. Although coaxial cables have wideband capabilities from d.c. up to well into the microwave band, attenuation increases with frequency and so coaxial cables may be designed to operate over definite frequency bands only such as audio, radio or video. However, their large bandwidth can be exploited for multichannel operation whereby several frequencies are sent down the same coaxial cable.

Coaxial cables normally use the TEM mode of propagation as shown in Fig. 1.2. To ensure that other modes do not exist, the cable size has to be decreased as the frequency increases, thus reducing its power handling capacity.

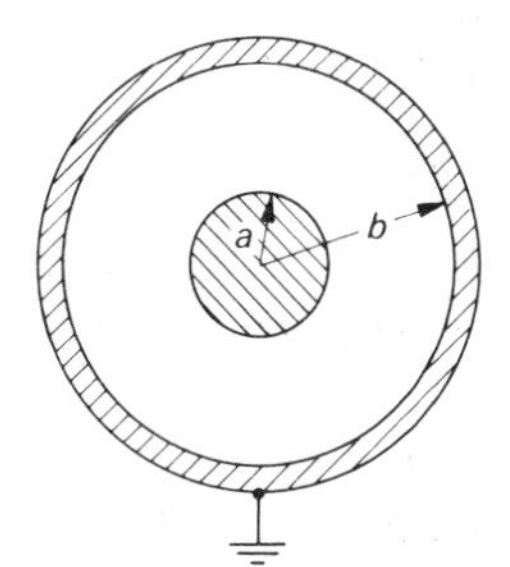

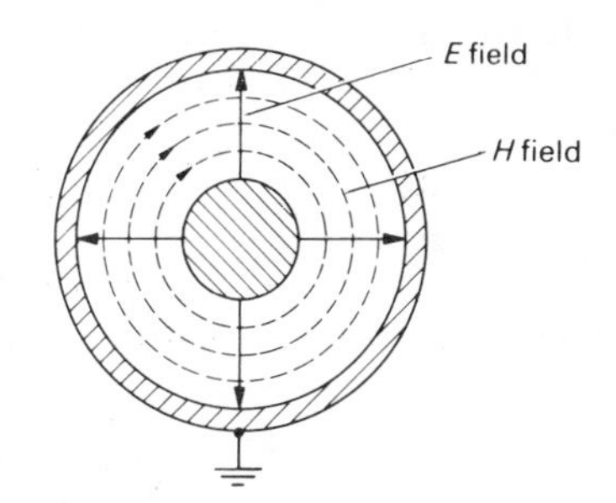

Fig. 1.2

A typical coaxial cable is characterised by R, L, G and C per unit length and its relevant parameters are given by:

$$R_{\text{a.c.}} = \frac{1}{2\pi}\left(\frac{\omega\mu\rho}{2}\right)^{1/2}\left[\frac{1}{a}+\frac{1}{b}\right] \text{ohm/m}$$

$$L = \frac{\mu_0\mu_r}{2\pi}\ln b/a \text{ henry/m}$$

$$G = \omega C \tan\delta \text{ siemens/m}$$

$$C = \frac{2\pi\varepsilon_0\varepsilon_r}{\ln b/a} \text{ farad/m}$$

$$Z_0 = \sqrt{L/C} = \frac{138}{\sqrt{\varepsilon_r}} \log_{10} b/a \text{ ohm}$$

$$v = \frac{1}{\sqrt{\mu\varepsilon}} = \frac{3 \times 10^8}{\sqrt{\mu_r \varepsilon_r}} \text{ m/s}$$

$$\alpha \simeq R/2Z_0 \text{ nepers/m}$$

where b is the inner radius of the outer conductor and a is the outer radius of the inner conductor, ρ is the specific resistance and δ is the loss angle of the capacitance C.

1.3 Stripline and microstrip[5–8]

A form of transmission line having low losses has been known for a long time. It is the parallel plate line with infinite plates and propagating a TEM wave. Such a system is only theoretical and not practical because of its infinite size and the difficulty of supporting the plates.

However, a form of line which uses finite plates and an intervening medium to support the plates has found growing importance recently. Two possible configurations exist and are known as *stripline* and *microstrip* as illustrated in Figs 1.3 and 1.5 respectively.

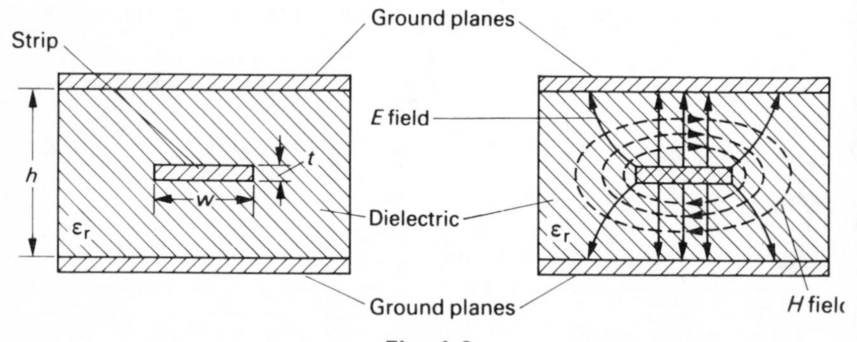

Fig. 1.3

Stripline consists essentially of two parallel 'ground planes' spaced a distance h apart and with a central strip conductor of width w and thickness t. It is centrally positioned within dielectric material with a relative permittivity of about 2 or 3. Propagation in stripline is somewhat similar to that in coaxial cable wherein the fields are confined between the outer and inner conductors. Thus, if the distance between the plates is small compared to a wavelength, losses are low due to the absence of higher-order modes, and propagation is essentially in the TEM mode with the electric and magnetic field configurations shown in Fig. 1.3.

Stripline parameters may be evaluated using conformal transformation or computerised techniques and yield results for the characteristic impedance Z_0, wavelength of propagation λ_g and attenuation coefficient α. By assuming zero thickness stripline for convenience, the characteristic impedance is given by the following expressions according to Cohn:

$$Z_0 = \frac{30}{\sqrt{\varepsilon_r}} \ln 2\left[\frac{1+\sqrt{k}}{1-\sqrt{k}}\right] \text{ohm}$$

for $w/h \leqslant 0{\cdot}5$ and

$$Z_0 = \frac{30\pi^2}{\sqrt{\varepsilon_r}} \Big/ \ln 2[(1+\sqrt{k'})/(1-\sqrt{k'})] \text{ ohm}$$

for $w/h > 0{\cdot}5$, where ε_r is the relative permittivity of the dielectric, $k = \text{sech}(\pi w/2h)$ and $k' = \tanh(\pi w/2h)$.

Furthermore, for TEM propagation the wavelength of propagation in a dielectric medium of relative permittivity ε_r is given by

$$\lambda_g = \lambda_0/\sqrt{\varepsilon_r}$$

where λ_0 is the free space wavelength. Typical graphs of the quantity $Z_0\sqrt{\varepsilon_r}$ for various values of w/h are shown in Fig. 1.4.

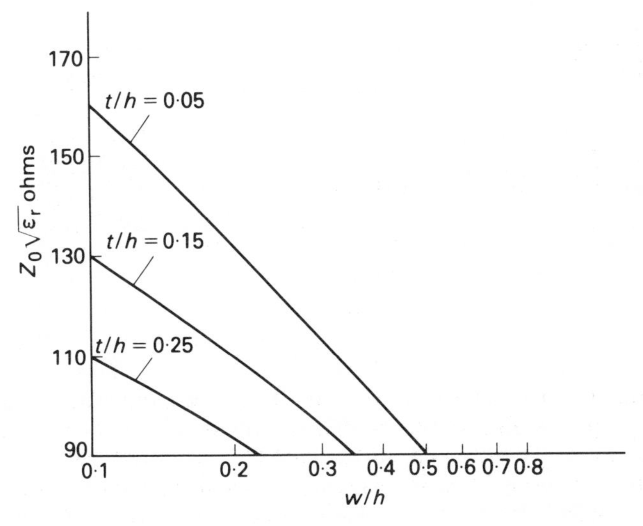

Fig. 1.4

Losses in stripline consist of conductor losses and dielectric losses. Since the latter are usually small at microwave frequencies, the attenuation coefficient α for a given value of characteristic impedance Z_0 is given by

$$\alpha = 4{\cdot}34 \frac{R_s}{Z_0 w} \text{ dB/cm}$$

where R_s is the skin resistivity of the conducting surfaces and w is the strip width in centimetres.

The alternative microstrip geometry consists of a narrow conductor which is supported on a dielectric substrate and mounted on a 'ground plane' as shown in Fig. 1.5. The structure may or may not be covered over by a metal enclosure serving as an earth shield. Because of the composite nature of the dielectric interface, propagation cannot be true TEM and for the complex field pattern shown in Fig. 1.5, it is considered to be *quasi-TEM*.

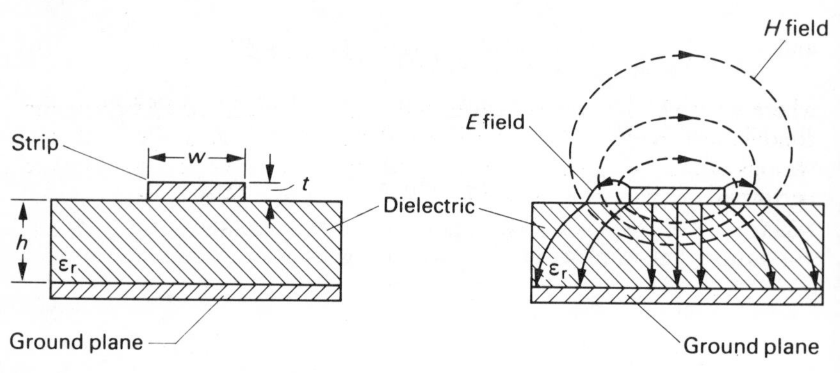

Fig. 1.5

Losses are generally greater in microstrip than in stripline but they can be reduced by using higher dielectric constant materials. Microstrip lines are usually fabricated on fibreglass or polystyrene printed circuit boards about 1·5 mm thick with copper conductors 3 mm wide. For integrated circuits, alumina, silicon or sapphire about 0·25 mm thick are used as substrates with conductors made of copper, aluminium or gold about 0·25 mm wide. More recently, circuits are also being manufactured with high-resistivity gallium arsenide. By using very thin substrates such as 0·1 mm quartz for example, the frequency limit of microstrip can be extended to about 100 GHz.

The characteristics of microstrip are difficult to evaluate exactly but depend on the ratio w/h and the dielectric constant ε_r. The numerical analysis may be undertaken using conformal transformation or computerised techniques to achieve the desired results. For typical microstrip lines with finite strip thickness, the characteristic impedance Z_0 is given by the following expressions according to Hammerstad:

$$Z_0 \simeq \frac{60}{\sqrt{\varepsilon_{\text{eff}}}} \ln\left(\frac{8h}{w}\right) \qquad (w/h \leqslant 1)$$

and

$$Z_0 \simeq \frac{377}{\sqrt{\varepsilon_{\text{eff}}}} \left(\frac{h}{w}\right) \qquad (w/h > 1)$$

where ε_{eff} is the effective dielectric constant of the substrate material. It is usually less than the value ε_r for the substrate and it takes account of the effect of the external fields.

The corresponding wavelength of propagation λ_g and attenuation coefficient α assume the following values respectively:

$$\lambda_g = \lambda_0/\sqrt{\varepsilon_{\text{eff}}}$$

and

$$\alpha \simeq 8{\cdot}68 \frac{R_s}{Z_0 w} \text{ dB/cm} \qquad (w/h > 1)$$

where λ_0 is the free space wavelength, R_s is the skin resistivity of the conducting surfaces and w is the strip width in cm. Typical graphs of Z_0 for various values of w/h and ε_r are shown in Fig. 1.6. Further details about the performance of stripline and microstrip are given in Appendix A.

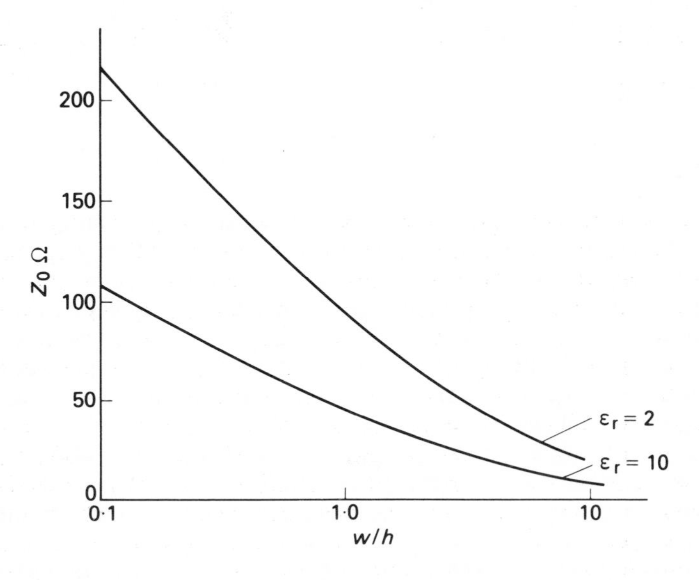

Fig. 1.6

1.4 Waveguides[9, 10]

For minimum losses and high-power transmission at microwave frequencies, the waveguide is used extensively. It consists essentially of a single metallic conductor in the shape of a rectangular box or hollow cylinder through which electromagnetic waves are propagated and is shown in Fig. 1.7. Such guided waves have field configurations somewhat different from those of the previous transmission lines considered and are called either transverse electric (TE) or transverse magnetic (TM) waves. Alternatively, they are known as *H* waves or *E* waves respectively.

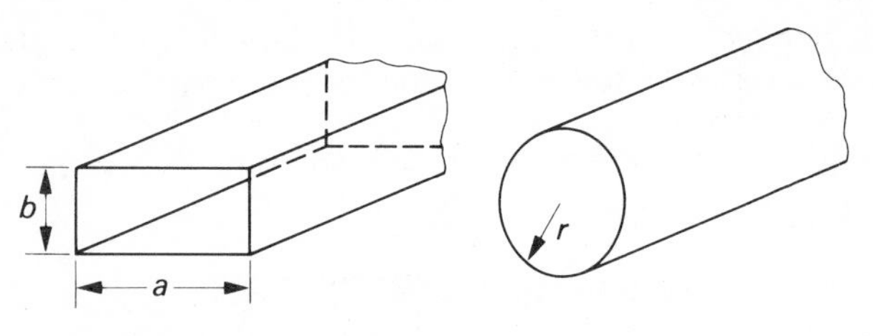

Fig. 1.7

Rectangular waveguides have fewer losses than circular waveguides and are less prone to mode changing. Hence, they are in most common use. Recently, however, the extremely low loss of one of the circular waveguide modes, the TE_{01} (H_{01}) wave, has been exploited for long distance communication. Previous problems, associated with mode changing and distortion due to phase change along the guide, are being overcome by the use of mode filters or by employing a helical waveguide. Furthermore, waveguide systems provide wide band capabilities and will continue to be used to meet the demands for multichannel circuits.

Waveguides are usually made of brass, copper or aluminium in various standard sizes corresponding to the frequencies used. To reduce losses, the inner walls are sometimes coated with a thin layer of silver or gold. Typical standard sizes for rectangular waveguides are given in Table 1.1.

At the lower microwave frequencies, hard drawn waveguides are usually manufactured, but at millimetric wavelengths special electroforming techniques are used for high precision. Waveguide sections are usually coupled together by flanged assemblies which are bolted together and supported periodically on metal stands.

The parameters associated with the rectangular or circular waveguide may be easily derived from theory using Maxwell's equations and are well established in practice. The important parameters of waveguide impedance Z_{TE} or Z_{TM} and cut-off wavelength λ_c are closely related to the dimensions of the waveguide, while the attenuation losses depend on these factors as well as on the inner surface finish and metal of the waveguide walls. Typical

Table 1.1

Waveguide designation	*Internal dimensions* (cm)	*Frequency range* (GHz)	*UK band*	*US band**
6	16·51 × 8·26	1·12 – 1·70	L	D
10	7·21 × 3·40	2·60 – 3·95	S	E/F
12	4·75 × 2·21	3·95 – 5·85	C	G
16	2·29 × 1·02	8·20 – 12·40	X	I/J
18	1·58 × 0·79	12·40 – 18·0	J	J
22	0·71 × 0·36	26·5 – 40·0	Q	K
26	0·31 × 0·15	60·0 – 90·0	O	M
		40·0 – 300	Mm	Mm

* Military designation.

values of λ_c for various rectangular and circular waveguide modes are given in Table 1.2. For rectangular waveguides, the values are related to the width a and height b of the waveguide. For circular waveguides, the values are related to the radius r of the waveguide.

Table 1.2

Rectangular modes	λ_c	*Circular modes*	λ_c
$TE_{10}(H_{10})$	$2a$	$TE_{01}(H_{01})$	$1{\cdot}64r$
$TE_{20}(H_{20})$	a	$TE_{11}(H_{11})$	$3{\cdot}42r$
$TE_{11}(H_{11})$	$0{\cdot}89a$	$TM_{01}(E_{01})$	$2{\cdot}61r$
$TM_{11}(E_{11})$	$0{\cdot}89a$	$TM_{11}(E_{11})$	$1{\cdot}64r$
$TM_{12}(E_{12})$	$0{\cdot}49a$	$TM_{12}(E_{12})$	$0{\cdot}896r$

For certain specific applications, other shapes are used such as square waveguides, ridged waveguides and flexible waveguides. Square waveguides with dimensions $a = b$, are useful for carrying circularly polarised waves consisting of horizontal and vertical polarisations. Rectangular waveguides consisting of a single or double ridge have a lower cut-off wavelength and this enables a smaller waveguide to be used at a given frequency. There is also an increase in the usable frequency range of the waveguide but with higher attenuation. Flexible waveguides are usually employed in small lengths only and are used between rigid sections to allow some movement between them or for awkward joints around corners or bends.

1.5 Optical fibres[11–13]

The transmission of light by means of optical fibres has led to the present development of optical communications over short and long distance telephone routes of the order of several kilometres. An optical fibre consists essentially of a central core of very pure silica glass with a high refractive index and a diameter of about 5–6 μm. It is surrounded by cladding material with a lower refractive index and an outer covering for mechanical strength or protection. It is known as a *monomode* fibre and attenuation losses as low as 0·5 dB/km are now being achieved. An alternative type of optical fibre with an attenuation around 4–5 dB/km, when installed in the ground, consists of a larger core between 20 μm and 150 μm, with the outer material having an abrupt or gradual change in the refractive index. It is known as a *multimode* fibre and the various types are illustrated in Fig. 1.8.

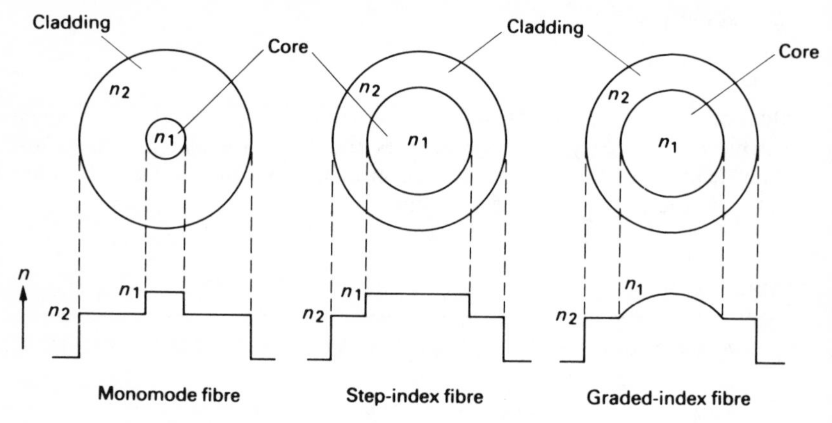

Fig. 1.8

Modulated light, at wavelengths usually in the near infra-red region around 700–900 nm when propagated down multimode fibres, is guided along by total reflection at the inner cladding surface (step-index type) or is continuously deflected inwards towards the axis of the core (graded-index type). The field patterns or mode patterns of the optical propagation usually have a component of both the electric field and the magnetic field in the direction of propagation. Hence, many modes can propagate at the same time giving rise to modal *dispersion*. In monomode fibres, however, owing to the small core diameter involved, propagation is essentially along the core centre by a single mode and a common mode of propagation is the HE_{11} mode. Problems of dispersion are therefore avoided but mechanical misalignment problems due to the small core size can lead to high attenuation losses.

The present rapid development and use of optical fibres for communication purposes is due to its several advantages, amongst which are wide bandwidth and relatively low transmission loss compared with cables. Moreover, optical fibre systems provide good electrical isolation between transmitter and receiver and they are immune to crosstalk or electromagnetic interference. Furthermore, the present progress has been spurred on by the advances made in the area of light sources such as lasers and light-emitting diodes (LEDs), together with the availability of solid-state optical receivers such as the PIN photodiode and the avalanche photodiode. The present state-of-the-art is such that practical communication systems are now being installed in various countries throughout the world.

2
Line transmission

Transmission lines used for communication purposes must operate over a range of frequencies and their behaviour is analysed in terms of resistance R, inductance L, conductance G and capacitance C, all defined per unit length of line. For the purposes of analysis, a line of infinite length may be considered and is known as an infinite line. Results obtained for such an infinite line can then be easily applied to the shorter, general line.

2.1 The general line

Consider the general two-wire line shown in Fig. 2.1 with primary line constants R, L, G and C defined per loop metre (one metre along each wire). If the voltage and current at the input to a short section of length δx are v and i, respectively, there is a drop of voltage across the section and leakage current between the lines. The corresponding values of voltage and current at the output of the section are given by using partial differentiation since v and i are alternating quantities and also vary with distance x along the line.

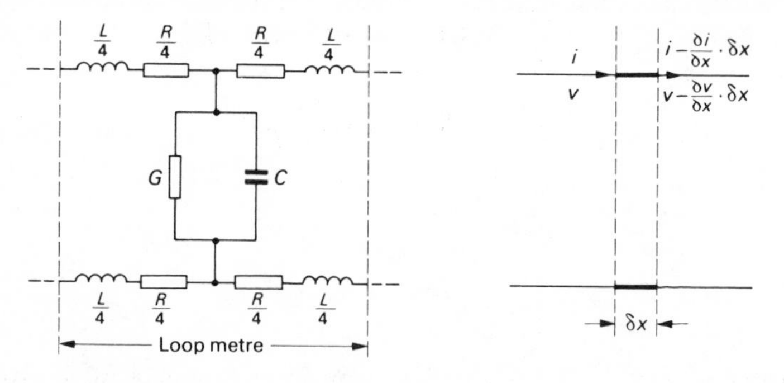

Fig. 2.1

Hence, we obtain the equations

$$-\frac{\partial v}{\partial x}\,\delta x = R\,\delta x i + L\,\delta x\frac{\partial i}{\partial t}$$

$$-\frac{\partial i}{\partial x}\,\delta x = G\,\delta x v + C\,\delta x\frac{\partial v}{\partial t}$$

or
$$-\frac{\partial v}{\partial x} = Ri + L\frac{\partial i}{\partial t} \tag{1}$$

$$-\frac{\partial i}{\partial x} = Gv + C\frac{\partial v}{\partial t} \tag{2}$$

A single sinusoidal voltage or current can be represented by the phasors

$$v = V\mathrm{e}^{\mathrm{j}\omega t}$$

$$i = I\mathrm{e}^{\mathrm{j}\omega t}$$

where V, I are functions of x only. Hence

$$\frac{\partial v}{\partial x} = \frac{\mathrm{d}V}{\mathrm{d}x}\mathrm{e}^{\mathrm{j}\omega t} \qquad \frac{\partial i}{\partial x} = \frac{\mathrm{d}I}{\mathrm{d}x}\mathrm{e}^{\mathrm{j}\omega t}$$

$$\frac{\partial v}{\partial t} = \mathrm{j}\omega V\mathrm{e}^{\mathrm{j}\omega t} \qquad \frac{\partial i}{\partial t} = \mathrm{j}\omega I\mathrm{e}^{\mathrm{j}\omega t}$$

Substituting these results in equations (1) and (2) yields

$$-\frac{\mathrm{d}V}{\mathrm{d}x} = (R + \mathrm{j}\omega L)I$$

$$-\frac{\mathrm{d}I}{\mathrm{d}x} = (G + \mathrm{j}\omega C)\,V$$

and on further differentiation we obtain

$$\frac{\mathrm{d}^2V}{\mathrm{d}x^2} = (R + \mathrm{j}\omega L)\,(G + \mathrm{j}\omega C)\,V$$

$$\frac{\mathrm{d}^2I}{\mathrm{d}x^2} = (R + \mathrm{j}\omega L)\,(G + \mathrm{j}\omega C)I$$

or
$$\frac{\mathrm{d}^2V}{\mathrm{d}x^2} = \gamma^2 V \tag{3}$$

$$\frac{\mathrm{d}^2I}{\mathrm{d}x^2} = \gamma^2 I \tag{4}$$

where $\gamma = \sqrt{(R + \mathrm{j}\omega L)\,(G + \mathrm{j}\omega C)}$ is complex and is called the propagation coefficient. It may be represented by

$$\gamma = \alpha + \mathrm{j}\beta$$

where α is the attenuation coefficient and β is the phase-change coefficient.

It is easily shown that

$$\alpha^2 - \beta^2 = RG - \omega^2 LC$$

and
$$2\alpha\beta = \omega\,(LG + RC)$$

2.2 Secondary line constants

The attenuation constant α accounts for the loss of voltage down the line, while β produces a regular phase shift along the line. This is so, because the wave requires finite time to travel down the line. The quantities α and β are called the secondary constants but are not really constant as they vary with frequency. A further quantity of interest is the velocity with which energy travels down the line and it is directly related to ω and β.

Consider the wave shown in Fig. 2.2 as travelling down the line with velocity v and angular frequency ω. The distribution of voltage over a distance $x = \lambda$, will correspond to one full cycle of variation as shown in Fig. 2.2 and the corresponding phase shift is equal to 2π radians.

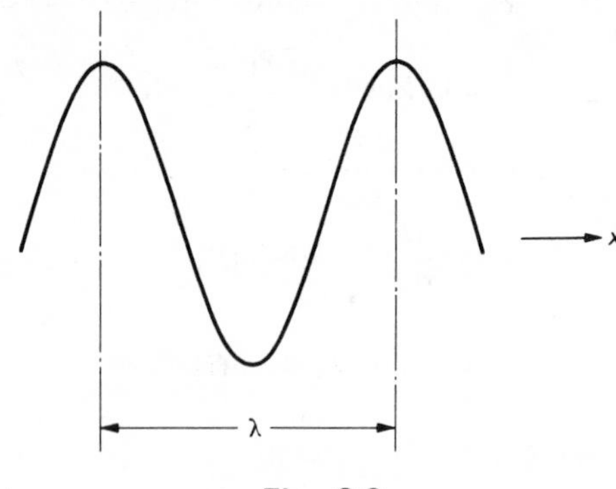

Fig. 2.2

Hence
$$\beta\lambda = 2\pi$$
or
$$\beta = \frac{2\pi}{\lambda}$$

Now $v = f\lambda$ where f is the frequency of operation. Hence

$$v = f\,\frac{2\pi}{\beta} = \frac{\omega}{\beta} \text{ metres/s}$$

Equations (3) and (4) above are second-order differential equations. They are of standard form and by differentiation and substitution, the solutions are readily shown to be

$$V = Ae^{-\gamma x} + Be^{\gamma x}$$

$$I = \frac{A}{Z_0}\,e^{-\gamma x} - \frac{B}{Z_0}\,e^{\gamma x}$$

where $Z_0 = \sqrt{(R + j\omega L)/(G + j\omega C)}$ is called the characteristic impedance of the line and A, B are constants determined from the boundary conditions.

2.3 Infinite line

For the infinite line shown in Fig. 2.3, let the sending end voltage and current be V_S, I_S respectively.

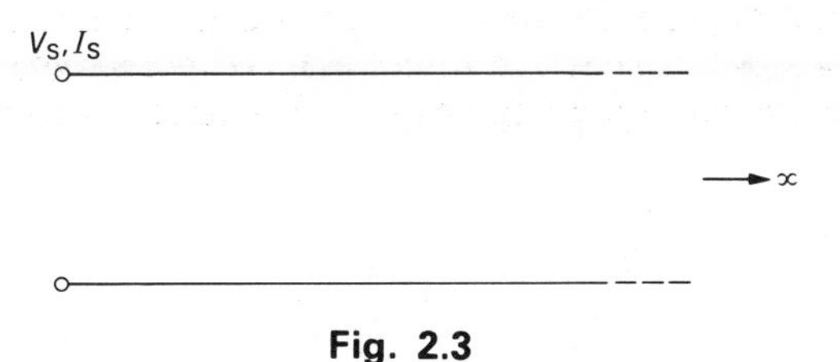

Fig. 2.3

At $x = 0$, we have

$$V = V_S = Ae^0 + Be^0$$

or

$$V_S = A + B$$

As x tends to infinity, $V \to 0$ since the line voltage is completely attenuated.

Hence

$$0 = Ae^{-\gamma x} + Be^{\gamma x}$$

with $Ae^{-\gamma x} \to 0$ as x tends to infinity and the only possible solution is $B = 0$ with $V_S = A + 0 = A$.

Hence

$$V = V_S e^{-\gamma x}$$

$$I = \frac{V_S}{Z_0} e^{-\gamma x}$$

as the corresponding equations for an infinite line. This leads to the evaluation of Z_0 as

$$\frac{V}{I} = \frac{V_S e^{-\gamma x}}{(V_S/Z_0)e^{-\gamma x}} = Z_0$$

or the ratio of voltage to current at the input to an infinite line or at any point in it is Z_0 and is given the name of characteristic impedance. It depends largely on R, L, G and C which are fixed by the design of a particular line. However, Z_0 also varies with f, the frequency of the wave on the line.

2.4 Hyperbolic solutions

Alternative solutions to the line equations use hyperbolic functions and yield a form useful for solving numerical problems.

We have

$$V(x) = Ae^{-\gamma x} + Be^{\gamma x}$$

$$I(x) = \frac{A}{Z_0} e^{-\gamma x} - \frac{B}{Z_0} e^{\gamma x}$$

Let V_S, I_S, V_R and I_R be the corresponding voltages and currents at the sending end and receiving end respectively.

At $x = 0$, $V = V_S$ and $I = I_S$ with

$$V_S = A + B$$

$$I_S = \frac{A}{Z_0} - \frac{B}{Z_0}$$

or

$$I_S Z_0 = A - B$$

giving

$$A = \frac{V_S + I_S Z_0}{2}$$

$$B = \frac{V_S - I_S Z_0}{2}$$

Substituting into $V(x)$ and $I(x)$ above yields

$$V(x) = e^{-\gamma x}\left[\frac{V_S + I_S Z_0}{2}\right] + e^{\gamma x}\left[\frac{V_S - I_S Z_0}{2}\right]$$

$$I(x) = \frac{e^{-\gamma x}}{Z_0}\left[\frac{V_S + I_S Z_0}{2}\right] - \frac{e^{\gamma x}}{Z_0}\left[\frac{V_S - I_S Z_0}{2}\right]$$

Since

$$\cosh \gamma x = \frac{e^{\gamma x} + e^{-\gamma x}}{2}$$

$$\sinh \gamma x = \frac{e^{\gamma x} - e^{-\gamma x}}{2}$$

Hence

$$V(x) = V_S \cosh \gamma x - I_S Z_0 \sinh \gamma x$$

$$I(x) = I_S \cosh \gamma x - \frac{V_S}{Z_0} \sinh \gamma x$$

2.5 Practical line

Since the input impedance of an infinite line is a constant and equal to Z_0, shorter practical lines using this property can be employed instead. This is illustrated in Fig. 2.4 where a short length l of an infinite line is cut away and the remainder which is still an infinite line whose input impedance is Z_0, is replaced by a lumped component of value Z_0, as shown on the right.

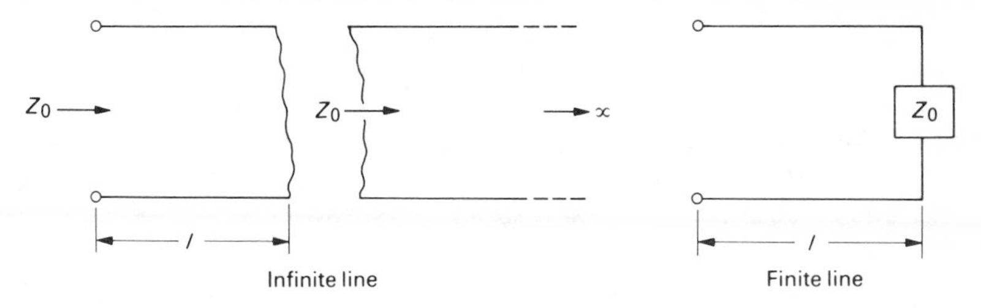

Fig. 2.4

A finite length of line which is terminated in Z_0 is called a correctly terminated or matched line and it has properties similar to an infinite line. Hence, the equations derived for the infinite line also hold in this case.

2.6 General termination

In many cases, the practical line may have a general termination Z_R and it is necessary to evaluate the input impedance Z_S.

Let the input impedance of the line shown in Fig. 2.5 be Z_S. Using the quantities given in Fig. 2.5 we obtain

$$Z_S = V_S/I_S$$

$$Z_R = V_R/I_R$$

At $x = l$

$$V_R = V_S \cosh \gamma l - I_S Z_0 \sinh \gamma l$$

$$I_R = I_S \cosh \gamma l - V_S/Z_0 \sinh \gamma l$$

or

$$Z_R = \frac{V_R}{I_R} = \frac{V_S \cosh \gamma l - I_S Z_0 \sinh \gamma l}{I_S \cosh \gamma l - V_S/Z_0 \sinh \gamma l}$$

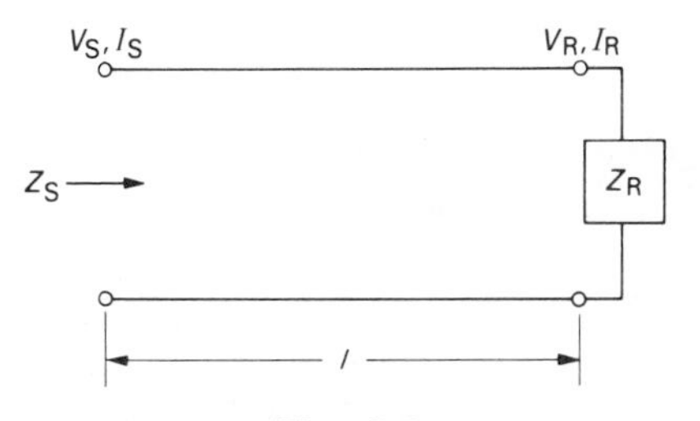

Fig. 2.5

Substituting for V_S from above, in terms of Z_S and I_S yields

$$\frac{V_S}{I_S} = Z_S = Z_0\left[\frac{Z_R \cosh \gamma l + Z_0 \sinh \gamma l}{Z_0 \cosh \gamma l + Z_R \sinh \gamma l}\right]$$

or

$$Z_S = Z_0\left[\frac{Z_R/Z_0 + \tanh \gamma l}{1 + Z_R/Z_0 \tanh \gamma l}\right]$$

which shows that Z_S is in general complex.

For high frequency lines operating around 100 MHz, α is small compared to β and so $\cosh \gamma l \simeq \cos \beta l$ and $\sinh \gamma l \simeq \mathrm{j} \sin \beta l$.

Hence

$$Z_S = Z_0\left[\frac{Z_R \cos \beta l + \mathrm{j}Z_0 \sin \beta l}{Z_0 \cos \beta l + \mathrm{j}Z_R \sin \beta l}\right]$$

or

$$Z_S = Z_0\left[\frac{Z_R + \mathrm{j}Z_0 \tan \beta l}{Z_0 + \mathrm{j}Z_R \tan \beta l}\right]$$

2.7 Special cases

Two special cases are the open-circuit line impedance and short-circuit line impedance, for the line of length l shown in Fig. 2.6.

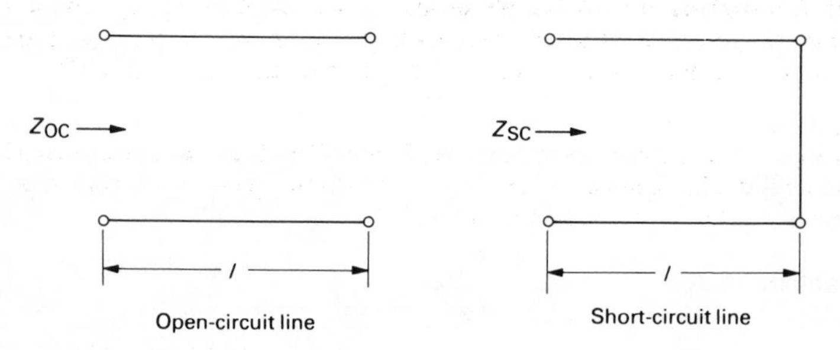

Fig. 2.6

(a) Open-circuit line

Here $Z_R = \infty$, $Z_S = Z_{OC}$. Hence

$$Z_{OC} = Z_0\left[\frac{\cosh \gamma l + Z_0/Z_R \sinh \gamma l}{Z_0/Z_R \cosh \gamma l + \sinh \gamma l}\right]$$

or

$$Z_{OC} = Z_0 \cosh \gamma l$$

(b) Short-circuit line

Here $Z_R = 0$, $Z_S = Z_{SC}$. Hence

$$Z_{SC} = Z_0\left[\frac{Z_R \cosh \gamma l + Z_0 \sinh \gamma l}{Z_0 \cosh \gamma l + Z_R \sinh \gamma l}\right]$$

or

$$Z_{SC} = Z_0 \tanh \gamma l$$

Two further equations can be obtained from the last two results. These are:

$$Z_0 = \sqrt{Z_{OC} Z_{SC}}$$

and

$$\tanh \gamma l = \sqrt{\frac{Z_{SC}}{Z_{OC}}}$$

from which Z_0 and γ for a line can be obtained.

Example 2.1

Derive an expression for the input impedance of a loss-free transmission line of length l, terminated by an impedance Z. Assume the voltage and current at distance x from the receiving end to be given by

$$V = Ae^{j\beta x} + Be^{-j\beta x}$$
$$IZ_0 = Ae^{j\beta x} - Be^{-j\beta x}$$

where A and B are constants, $\beta = \omega\sqrt{LC}$ is the phase-change coefficient and $Z_0 = \sqrt{L/C}$ is the characteristic impedance, L and C are the inductance and capacitance per unit length of line and ω is the angular frequency.

If $L = 0{\cdot}60\ \mu\text{H/m}$, $C = 240\ \text{pF/m}$ and $\omega = 2\pi \times 10^8$ rad/s, determine (1) the phase-change coefficient β and wavelength λ in the line and (2) the input impedance for a line of length $l = \lambda/4$ terminated by an impedance $Z = -j100\,\Omega$. (U.L.)

Solution

The input impedance has been derived in Section 2.6 where x was measured from the sending end. The expressions for V and I given in the question will correspond with those in the text if x is replaced by $-x$.

Problem

$$\beta = \omega\sqrt{LC}$$
$$= 2\pi \times 10^8 \sqrt{6 \times 10^{-7} \times 240 \times 10^{-12}}$$

or

$$\beta = 2{\cdot}4\pi$$

Also

$$\lambda = \frac{2\pi}{\beta} = \frac{2\pi}{2{\cdot}4\pi} = 0{\cdot}833\ \text{m}$$

and

$$Z_0 = \sqrt{\frac{L}{C}} = \left[\frac{6 \times 10^{-7}}{240 \times 10^{-12}}\right]^{1/2} = 50\,\Omega$$

Now

$$Z_{in} = Z_0\left[\frac{Z_R \cos \beta l + jZ_0 \sin \beta l}{Z_0 \cos \beta l + jZ_R \sin \beta l}\right]$$

where $$Z_R = -j100$$

$$\beta l = \frac{2\pi}{\lambda}\cdot\frac{\lambda}{4} = \frac{\pi}{2}$$

Hence $$\cos\beta l = \cos\pi/2 = 0$$

$$\sin\beta l = \sin\pi/2 = 1$$

with $$Z_{in} = Z_0\left[\frac{jZ_0}{jZ_R}\right] = Z_0^2/Z_R$$

$$= \frac{(50)^2}{-j100}$$

or $$Z_{in} = j25\,\Omega$$

2.8 Line classification

Transmission lines are used over a wide range of frequencies and have certain particular characteristics which need further consideration.

(a) Loss-free line

In this case $R = 0$, $G = 0$. Here

$$\gamma = \sqrt{j\omega L \times j\omega C} = j\omega\sqrt{LC}$$

Equating real and imaginary parts yields

$$\alpha = 0$$

$$\beta = \omega\sqrt{LC}$$

Also $$Z = \sqrt{\frac{j\omega L}{j\omega C}} = \sqrt{\frac{L}{C}}$$

which is a pure resistance.

This is a rather ideal case and the nearest practical example is the low-loss line.

(b) Low-loss line

Here $G \simeq 0$ and R is small where $R \ll \omega L$. Hence

$$\gamma = \sqrt{(R + j\omega L)(j\omega C)} = j\omega\sqrt{(R/j\omega + L)C}$$

$$= j\omega\sqrt{LC(1 + R/j\omega L)}$$

$$= j\omega\sqrt{LC}\left[1 - j\frac{R}{\omega L}\right]^{1/2}$$

$$= j\omega\sqrt{LC}\left[1 - j\frac{R}{2\omega L}\right]$$

by the Binomial theorem.

Hence $$\gamma = \frac{\omega R\sqrt{LC}}{2\omega L} + j\omega\sqrt{LC}$$

$$= \frac{R\sqrt{LC}}{2L} + j\omega\sqrt{LC}$$

$$= \frac{R}{2}\sqrt{\frac{C}{L}} + j\omega\sqrt{LC}$$

giving $$\alpha = \frac{R}{2}\sqrt{\frac{C}{L}}$$

$$\beta = \omega\sqrt{LC}$$

Also $$Z_0 = \sqrt{\frac{R + j\omega L}{j\omega C}} = \sqrt{\frac{L}{C}[1 - j(R/\omega L)]}$$

$$\simeq \sqrt{\frac{L}{C}}[1 - R/\omega l]^{1/2}$$

or $$Z_0 \simeq \sqrt{\frac{L}{C}}[1 - j(R/2\omega L)] \simeq \sqrt{\frac{L}{C}} - j\frac{R}{2\omega}\sqrt{\frac{1}{LC}}$$

This reduces to $\sqrt{L/C}$, a pure resistance at the higher frequencies, due to the large value of ω in the denominator of the second term.

Hence $$\alpha \simeq R/2Z_0$$

$$\beta = \omega\sqrt{LC}$$

If $G \neq 0$, then $$\alpha \simeq R/2Z_0 + GZ_0/2$$

(c) Low frequency line

Typically, it is an audio frequency telephone line and G and L may be low.

Hence $$j\omega L \ll R$$

$$G \ll j\omega C$$

with $$\gamma = \sqrt{R(j\omega C)}$$

$$= \sqrt{j\omega RC}$$

$$= \sqrt{\omega RC}\,\underline{/45^\circ}$$

or $$\gamma = \sqrt{\omega RC}\cos 45^\circ + j\sqrt{\omega RC}\sin 45^\circ$$

Hence $$\alpha = \beta = \sqrt{\frac{\omega RC}{2}} \quad \text{numerically}$$

Also $$v = \omega/\beta = \sqrt{\frac{2\omega}{RC}}$$

and $$Z_0 = \sqrt{\frac{R}{j\omega C}} = \sqrt{\frac{R}{\omega C}} \angle{-45^\circ}$$

(d) High frequency line

A two-wire radio-frequency line operating around 100 MHz comes under this classification.

Here $$\omega L \gg R$$
$$\omega C \gg G$$

Hence $$\gamma = \sqrt{j\omega L \times j\omega C} = j\omega\sqrt{LC}$$

giving $$\alpha = 0$$
$$\beta = \omega\sqrt{LC}$$

with $$v = \omega/\beta = 1/\sqrt{LC}$$

As L and C are very small for such lines v tends to c, the velocity of light which equals 3×10^8 m/s.

Also $$Z_0 = \sqrt{\frac{j\omega L}{j\omega C}} = \sqrt{L/C}$$

a pure resistance.

(e) Distortionless line

This type of line is of interest in designing practical lines with little or no distortion. The two types of distortion which are possible are attenuation distortion, in which certain frequencies are severely attenuated and may even disappear, and phase distortion, in which different frequencies suffer different amount of phase shift which is likely to be serious, especially in television.

The analysis is based on the fact that α and β are both functions of frequency, as is shown by the following expressions:

$$\alpha^2 = \tfrac{1}{2}[\{(RG+\omega^2 LC)^2 + \omega^2(LG-RC)^2\}^{1/2} + (RG-\omega^2 LC)]$$
$$\beta^2 = \tfrac{1}{2}[\{(RG+\omega^2 LC)^2 + \omega^2(LG-RC)^2\}^{1/2} - (RG-\omega^2 LC)]$$

Putting $$L/R = C/G \text{ or } LG = RC$$

yields

$$\alpha \simeq \sqrt{RG}$$
$$\beta \simeq \omega\sqrt{LC}$$
$$v = \omega/\beta = 1/\sqrt{LC} \qquad \text{(a constant)}$$

Hence, α is independent of frequency and β is proportional to frequency, the latter amounting to a linear phase characteristic.

The first condition $\alpha \simeq \sqrt{RG}$ yields no attenuation distortion, since all frequencies are attenuated by the same amount. The second condition $\beta \simeq \omega\sqrt{LC}$ yields no phase distortion since the velocity of travel of all frequencies is the same and their relative phase difference is unchanged down the line.

Also

$$Z_0 = \sqrt{\frac{R + j\omega L}{G + j\omega C}} = \sqrt{\frac{L(R/L + j\omega)}{C(G/C + j\omega)}}$$

or

$$Z_0 = \sqrt{L/C} \qquad \text{which is a pure resistance.}$$

The condition $LG = RC$ is called the *distortionless condition* and signifies the ideal condition for no attenuation or phase distortion. It is difficult to achieve in a practical line such as a cable since G is small. Hence L is increased artificially at intervals along the line by inserting lumped 'loading coils', a technique known as 'loading'. This practice, which was well established in earlier days, has now been superseded by the use of repeaters, both in underground and submarine cables.

The overall effect of loading was to produce a low-pass filter with a cut-off frequency which had a much lower α over the pass-band than its unloaded counterpart as shown in Fig. 2.7.

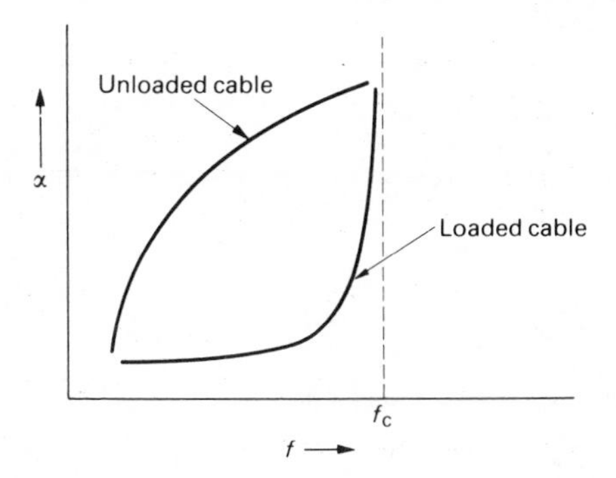

Fig. 2.7

2.9 Phase and group delay[14]

When a single frequency signal travels down a line, it suffers a phase shift of β rad/m. Hence, over a wavelength λ, the phase shift is 2π rads and we obtain

$$\beta\lambda = 2\pi$$

or

$$\beta = 2\pi/\lambda = 2\pi f / f\lambda = \omega/v_p$$

where v_p is defined as the phase velocity of the wave.

Hence

$$v_p = \omega/\beta \text{ metres/s.}$$

The reciprocal of v_p has the dimensions of seconds if the distance considered is a metre. It is called the phase delay, or time delay suffered by the wave in travelling unit distance.

Hence

$$\text{Phase Delay} = 1/v_p = \beta/\omega \text{ seconds}$$

For no phase distortion to occur, β must be proportioned to ω i.e. a linear phase characteristic. Hence, the corresponding phase delay must be constant and independent of frequency.

Similarly, when two or more frequencies are present on the line as in an A.M. signal, the waves combine to form a group and the peak of the envelope travels forward with a group velocity given by

$$v_g = \mathrm{d}\omega/\mathrm{d}\beta \text{ m/s}$$

The reciprocal of this also has the dimensions of seconds per unit distance and corresponds to a group delay.

Hence

$$\text{Group Delay} = 1/v_g = \mathrm{d}\beta/\mathrm{d}\omega \text{ seconds}$$

For the group of frequencies to remain in step as a group, the group delay must also be constant. Hence, $\mathrm{d}\beta/\mathrm{d}\omega$ must be constant, which amounts to a linear variation of β with ω over the range of frequencies of the group, otherwise distortion will again arise.

Phase or group delay is not serious in audio amplifiers or telephone lines provided it is not too great, as the ear cannot distinguish between small phase differences. However, in television, group delay is very important. It can lead to distortion since the picture brightness depends upon the peak of the video signal, which must not be out of step due to variable group delay. Figure 2.8 shows a typical plot of group delay in a video cable.

Example 2.2

A loss-free cable is a quarter of a wavelength long and is excited by a 1·0 V constant voltage sinusoidal source. The cable is terminated by a resistor of 73·5 Ω, which is not quite equal to the characteristic impedance of the cable. The input current is found to be 15·0 mA. Determine (1) the characteristic impedance of the cable; (2) the phase of the input current relative to that of the voltage source. (U.L.)

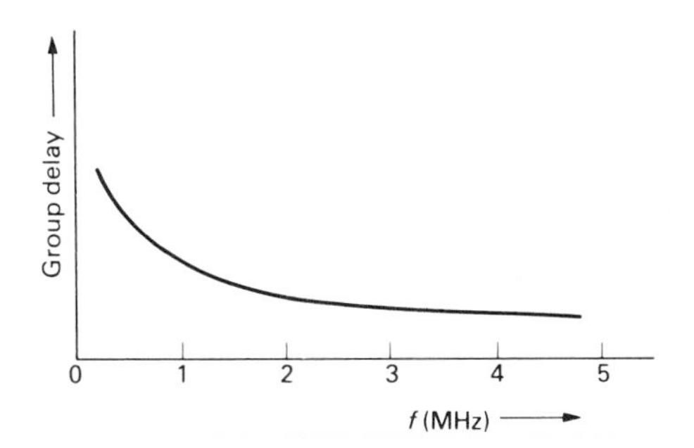

Fig. 2.8

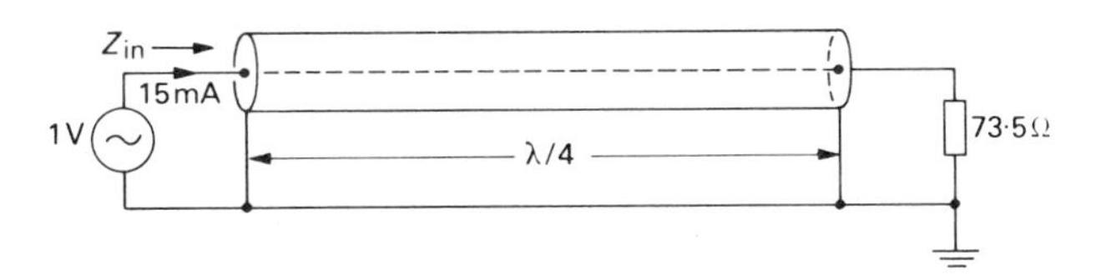

Fig. 2.9

Solution

$$Z_{in} = Z_0\left[\frac{Z_R \cosh \gamma l + Z_0 \sinh \gamma l}{Z_0 \cosh \gamma l + Z_R \sinh \gamma l}\right]$$

$$= Z_0\left[\frac{Z_R + Z_0 \tanh \gamma l}{Z_0 + Z_R \tanh \gamma l}\right]$$

Since the cable is loss free, $\alpha = 0$.

Hence $\tanh \gamma l = \dfrac{\sinh \gamma l}{\cosh \gamma l} = \dfrac{\mathrm{j} \sin \beta l}{\cos \beta l} = \mathrm{j} \tan \beta l$

with $Z_{\text{In}} = Z_0\left[\dfrac{Z_R + \mathrm{j}Z_0 \tan \beta l}{Z_0 + \mathrm{j}Z_R \tan \beta l}\right] = Z_0\left[\dfrac{Z_R/\tan \beta l + \mathrm{j}Z_0}{Z_0/\tan \beta l + \mathrm{j}Z_R}\right]$

Now $\tan \beta l = \tan(2\pi/\lambda . \lambda/4) = \tan \pi/2 = \infty$

Hence $Z_{in} = Z_0\left[\dfrac{0 + \mathrm{j}Z_0}{0 + \mathrm{j}Z_R}\right] = Z_0^2/Z_R$

with $Z_0^2 = Z_{in} Z_R = \dfrac{73{\cdot}5}{15 \times 10^{-3}}$

or $Z_0 = 70\,\Omega$ a pure resistance

Since Z_0 and Z_R are pure resistances, $Z_{in} = Z_0^2/Z_R$ is also a pure resistance. It is, in fact, the resistive load of the generator and so the input-voltage and current are in phase.

Example 2.3
Explain what is meant by the terms phase and group velocity.

Derive an expression for the phase coefficient of a transmission line in terms of the resistance R, inductance L, and capacitance C per unit length of line (the leakance is negligible). Hence, obtain expressions for the phase and group velocities in the line. Show that when the quantity $Q\ (=\omega L/R)$ is large both velocities approach a value of $1/\sqrt{LC}$. (U.L.)

Solution
The explanation of phase and group velocities is given in the text in Section 2.9. The propagation coefficient is given by

$$\begin{aligned}\gamma &= \sqrt{(R+\mathrm{j}\omega L)(G+\mathrm{j}\omega C)}\\ &= \sqrt{(R+\mathrm{j}\omega L)(\mathrm{j}\omega C)}\\ &= \mathrm{j}\omega\sqrt{(R/\mathrm{j}\omega+L)C}\\ &= \mathrm{j}\omega\sqrt{(R/\mathrm{j}\omega+1)LC} = \mathrm{j}\omega\sqrt{LC}[1-\mathrm{j}(R/\omega L)]^{1/2}\end{aligned}$$

Since Q is large, $\omega L \gg R$

Hence $\gamma = \mathrm{j}\omega\sqrt{LC}[1-\mathrm{j}(R/2\omega L)]$

by the Binomial theorem

or
$$\gamma = \frac{\omega R\sqrt{LC}}{2\omega L} + \mathrm{j}\omega\sqrt{LC}$$

Hence
$$\alpha = \frac{R}{2}\sqrt{C/L}$$

$$\beta = \omega\sqrt{LC}$$

Now
$$v_\mathrm{p} = \omega/\beta = \frac{\omega}{\omega\sqrt{LC}} = 1/\sqrt{LC}$$

$$v_\mathrm{g} = \mathrm{d}\omega/\mathrm{d}\beta = \frac{\mathrm{d}}{\mathrm{d}\beta}(\beta/\sqrt{LC}) = 1/\sqrt{LC}.$$

Hence, both v_p and v_g approach the value $1/\sqrt{LC}$ when Q is large.

2.10 Reflections on lines

If the terminating impedance on a transmission line is not Z_0, then a discontinuity exists at the end of the line and reflected waves of voltage and current will be set up on the line and they travel back towards the sending end. As far as energy is concerned, this means that some of the incident energy is reflected back, while the rest is absorbed in the load. The extreme cases of reflection occur on open-circuit and short-circuit.

(a) Open-circuited line

When the incident wave reaches the open circuit at the end of the line, the magnetic field collapses since the current is reduced to zero. This induces a voltage on the line which adds to the existing voltage on the line and equals it in magnitude. Hence, there is a *voltage doubling* effect.

The induced voltage on the line then travels back along the line and may be absorbed by the generator impedance if it equals Z_0, the characteristic impedance of the line. If not, it travels to and fro along the line till it is finally attenuated completely.

(b) Short-circuited line

Since the voltage is zero at the end of the line, there must be a phase reversal of the incident voltage at the end of the line so that the incident and reflected waves cancel one another at the end of the line. The reflected wave then travels back along the line and is either absorbed by the generator impedance or completely attenuated, as in the previous case.

Hence, in both these cases, there are two waves on the line, (a) the incident wave, and (b) the reflected wave. The resultant wave at any part of the line is the algebraic sum of the two waves and it produces a *standing wave* on the line.

The incident and reflected waves travel along the line and hence are called *travelling waves.* The resultant standing wave pattern on the line varies in magnitude, but remains *stationary* and contains maxima and minima at certain points on the line, where adjacent maxima or minima are separated by $\lambda/2$. A diagram illustrating these various points for two different instances of time is shown in Fig. 2.10 in which the line is assumed to be lossless and of length λ. The rms value of the voltage and current at various points on the line is shown in Fig. 2.11.

The ratio of voltage to current waves gives the impedance at the various points on the line. This is shown in Fig. 2.11 and for convenience, distances are measured from the receiving end. In particular, it will be seen that the input impedance of a $\lambda/4$ open-circuited line is zero, while that of a $\lambda/4$ short-circuited line is infinite.

2.11 Reflection coefficient

A reflection coefficient ρ at the receiving end is defined as the ratio of the reflected wave to the incident wave. At the end of a line of length l, the incident wave is $Ae^{-\gamma l}$ and the reflected wave $Be^{\gamma l}$. Hence

$$\rho = \frac{Be^{\gamma l}}{Ae^{-\gamma l}}$$

In general, ρ is complex and may be written as

$$\rho = |\rho| e^{j\phi}$$

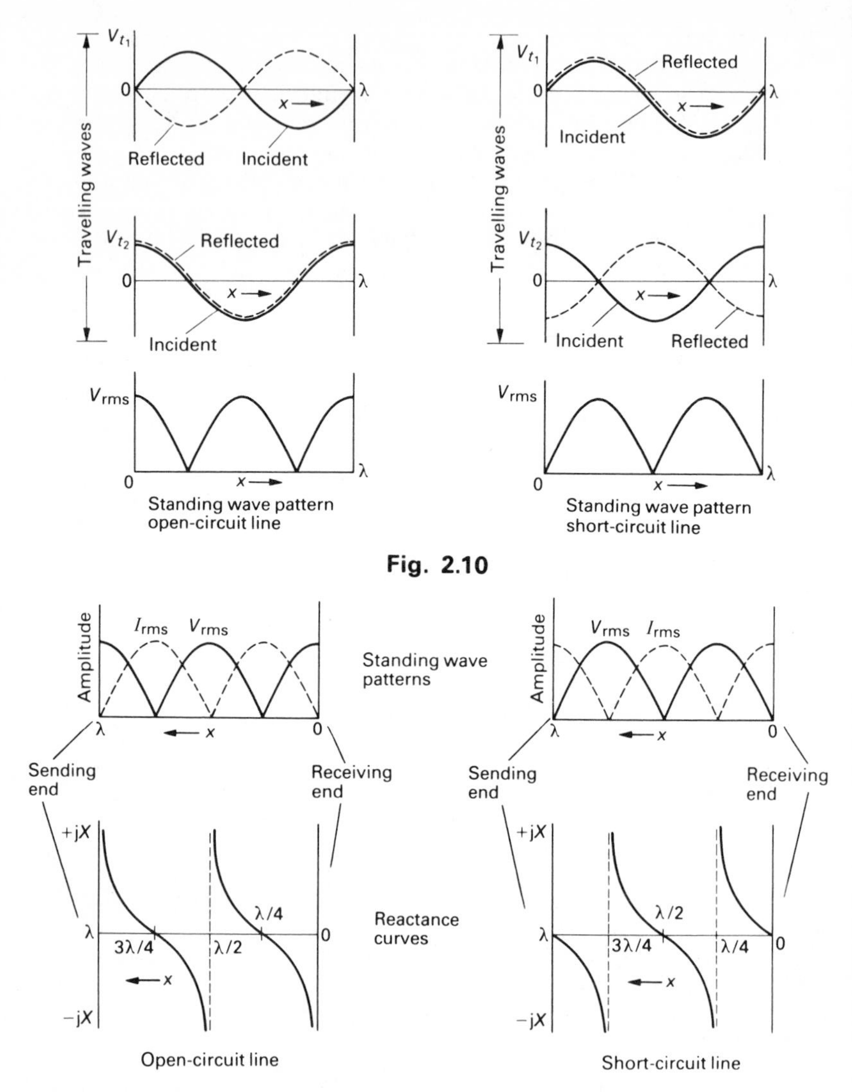

Fig. 2.10

Fig. 2.11

where $|\rho|$ is its magnitude and ϕ its phase angle. It is related to the load impedance Z_R and the characteristic impedance Z_0 as may be seen from below. For the general line equations we have at $x = l$,

$$V_R = Ae^{-\gamma l} + Be^{\gamma l} = Ae^{-\gamma l} + \rho Ae^{-\gamma l}$$

$$I_R = (A/Z_0)e^{-\gamma l} - (B/Z_0)e^{\gamma l} = (A/Z_0)e^{-\gamma l} - (\rho/Z_0)Ae^{-\gamma l}$$

or

$$V_R = Ae^{-\gamma l}(1+\rho)$$

$$Z_0 I_R = Ae^{-\gamma l}(1-\rho)$$

or

$$Z_R/Z_0 = \frac{1+\rho}{1-\rho}$$

Hence

$$\rho = \frac{Z_R - Z_0}{Z_R + Z_0} \quad \text{where} \quad 0 \leqslant \rho \leqslant 1$$

Comments

1. For an open-circuited line, $Z_R = \infty$. Hence

$$\rho = \frac{1 - Z_0/Z_R}{1 + Z_0/Z_R} = 1$$

2. For a short-circuited line, $Z_R = 0$. Hence

$$\rho = \frac{Z_R - Z_0}{Z_R + Z_0} = \frac{-Z_0}{Z_0} = -1$$

   and there is a phase reversal of voltage.

3. For a matched line, $Z_R = Z_0$. Hence

$$\rho = \frac{Z_0 - Z_0}{Z_0 + Z_0} = 0$$

   and there is no reflected wave.

2.12 Voltage standing wave ratio (VSWR)

When a transmission line is terminated in an arbitrary impedance $Z_R \neq Z_0$, an incident and a reflected wave are both present on the line. If A and B are their respective amplitudes at some point on the line, a maximum occurs when the two amplitudes are in phase and a minimum, when the two amplitudes are 180° out of phase.

Hence

$$|V_{max}| = A + B = A(1 + B/A)$$

$$|V_{min}| = A - B = A(1 - B/A)$$

The standing wave ratio s is defined as the ratio of $|V_{max}|$ to $|V_{min}|$. Hence

$$s = \frac{|V_{max}|}{|V_{min}|} = \frac{1 + B/A}{1 - B/A}$$

Now the reflection factor at the receiving end was defined as

$$\rho = Be^{\gamma l}/Ae^{-\gamma l}$$

with $$|\rho| = B/A$$

Substituting for B/A in the expression for s, we obtain

$$s = \frac{1+|\rho|}{1-|\rho|}$$

or
$$|\rho| = \frac{s-1}{s+1}$$

The standing wave ratio s is thus directly related to $|\rho|$ and since the latter is related to Z_0 and Z_R, for a purely *resistive* Z_0 and Z_R we obtain

$$s = \frac{1+|\rho|}{1-|\rho|} = \left[1+\left(\frac{Z_R - Z_0}{Z_R + Z_0}\right)\right] \Big/ \left[1-\left(\frac{Z_R - Z_0}{Z_R + Z_0}\right)\right] = \frac{Z_R}{Z_0}$$

or $$s = \frac{Z_R}{Z_0} \quad \text{if} \quad Z_R > Z_0$$

and $$s = \frac{Z_0}{Z_R} \quad \text{if} \quad Z_R < Z_0$$

Hence, measurements of s yield a direct knowledge of the nature of the load termination Z_R.

2.13 The Smith chart[15]

A convenient graphical way of solving transmission line problems is by means of the Smith chart, since it shows vividly, the variation of impedance along a transmission line. The chart consists of two sets of circles, one of which is a polar coordinate system with its centre as the chart centre at the point (1,0) and the other an orthogonal set of circles passing through the point denoted as $r = \infty$ on the chart.

In the first set of circles, the radial distance represents the magnitude of the reflection coefficient ρ and the angular distance its phase angle ϕ. The values of ρ vary from 0 to 1·0 and ϕ varies from 0 to 180°. These coordinates are not normally drawn to avoid overcrowding the chart and are shown dotted in Fig. 2.12(a) as an illustration. The coordinates also represent values of the standing wave ratio s and are called VSWR circles.

The second set of circles shown in full lines in Fig. 2.12(b) are orthogonal to one another and represent the resistance and reactive components of impedance whose values are all *normalised*. The normalised resistance r and reactance x are given in lower case letters by

$$r = R/Z_0$$

$$x = X/Z_0$$

where R, X are the actual values of resistance and reactance and r, x are the

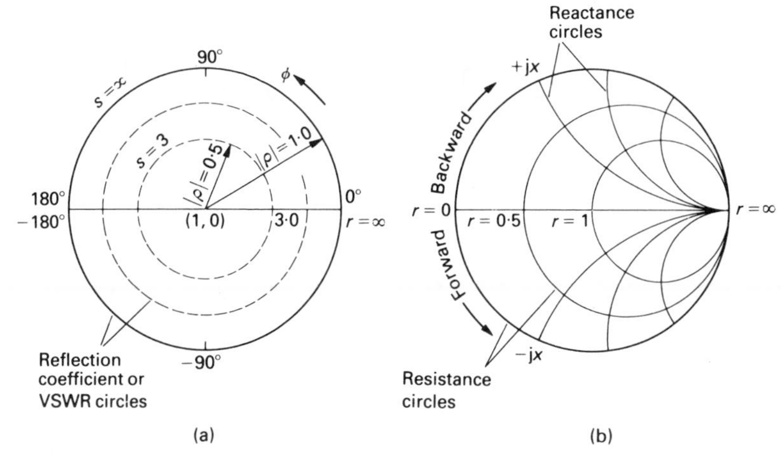

Fig. 2.12

corresponding normalised values which are obtained by dividing the actual values by Z_0, the characteristic impedance of the line. Both r and x range in values from 0 to ∞, the outermost circle of the chart corresponding to $r = 0$. The centres of these circles lie on two perpendicular lines and all the circles pass through the point $r = \infty$.

Circles of resistance are drawn in completely, while only part of the circles of reactance appears on the chart. The intersection of a resistance circle and a reactance circle gives a point of normalised impedance z and its inversion about the centre gives the normalised admittance y. Thus

$$z = Z/Z_0 = r \pm \mathrm{j}x$$

or

$$y = Y/Y_0 = g \pm \mathrm{j}b$$

The construction of these orthogonal circles can be obtained by a consideration of any load impedance Z_R. For example, let the normalised impedance be given by

$$z = Z_R/Z_0 = r + \mathrm{j}x \qquad \text{(assuming positive reactance only)}$$

and we also have

$$\frac{Z_R}{Z_0} = \frac{1+\rho}{1-\rho} \qquad \text{(from Section 2.11)}$$

with

$$z = r + \mathrm{j}x = (1+\rho)/(1-\rho)$$

Since ρ is complex, let $\rho = a + \mathrm{j}b$ and we obtain

$$z = r + \mathrm{j}x = \frac{1 + a + \mathrm{j}b}{1 - a - \mathrm{j}b}$$

Thus $$r+jx=\frac{(1+a)+jb}{(1-a)-jb}=\frac{(1-a^2-b^2)+2jb}{(1-a)^2+b^2}$$

after rationalising the denominator by the conjugate factor $(1-a+jb)$. By equating real and imaginary parts we then obtain

$$r=\frac{1-a^2-b^2}{(1-a)^2+b^2}$$

and $$x=\frac{2b}{(1-a)^2+b^2}$$

Now, expanding these expressions and re-arranging terms yields respectively

$$a^2-\frac{2ar}{(1+r)}+b^2=\frac{1-r}{1+r}$$

and $$a^2-2a+b^2-2b/x=-1$$

The two expressions above are the equations of circles and by the method of completing the squares we obtain

$$\left[a-\frac{r}{(1+r)}\right]^2+b^2=\frac{1}{(1+r)^2} \tag{1}$$

and $$(a-1)^2+(b-1/x)^2=1/x^2 \tag{2}$$

Equation (1) represents a set of resistance circles with centre at the point $[r/(1+r), 0]$ and radius $[1/(1+r)]$, while equation (2) represents a set of reactance circles with centre at the point $(1, 1/x)$ and radius $1/x$. Typical sets of resistance and reactance circles are illustrated in Fig. 2.13 for various values of r and x.

The final Smith chart is drawn by moving the x-ordinate by unity and using the relative chart centre (1, 0) instead of the actual centre point (0, 0). The reactance circles are then drawn in partially as stated earlier and this is illustrated in Fig. 2.14.

In Fig. 2.14, the outermost circle or 'base' circle is scaled off in wavelengths from 0 to $0{\cdot}5\lambda$ towards the generator i.e. in a backward direction or from 0 to $0{\cdot}5\lambda$ towards the load i.e. in a forward direction and they correspond to the *direction of movement* along the transmission line. Only half a wavelength is scaled off since the standing wave pattern repeats itself every $0{\cdot}5\lambda$ as is illustrated in Fig. 2.15.

Comments

1. This description of the Smith chart has assumed a lossless transmission line.
2. For a lossy line, the magnitude of the reflection coefficient will change along the line.

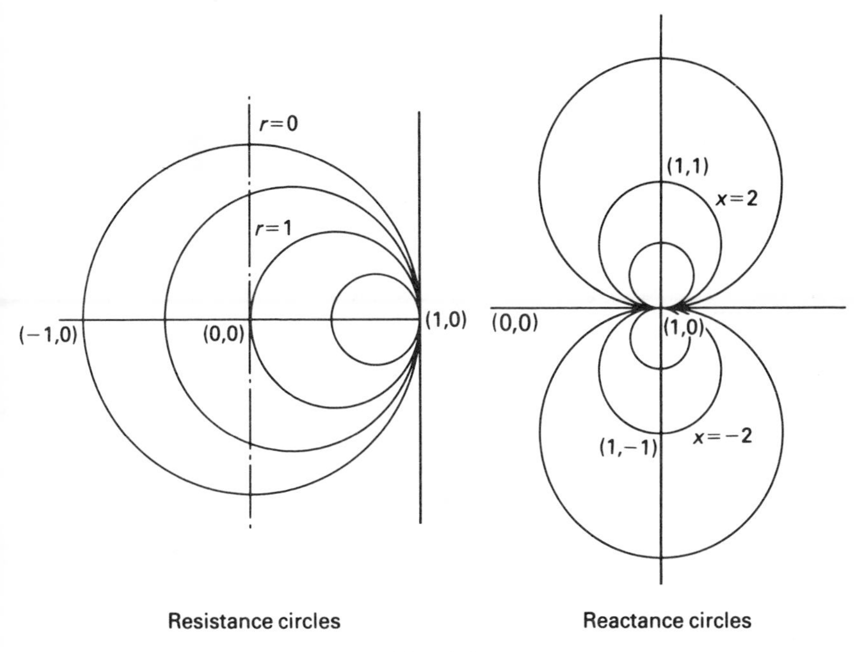

Fig. 2.13

3. For solving problems on lossy lines, other standard 'scales' may be used in conjunction with the Smith chart.

2.14 Typical examples

Load impedance

To plot an impedance $50+j100\,\Omega$ on a $50\,\Omega$ line.

1. Normalising Z_R yields $z_R = (50+j100)/50 = 1+j2{\cdot}0\,\Omega$.
2. Set chart with the central diameter horizontal and the point zero on the left. Starting from $r = 0$ on the central diameter move to point $r = 1{\cdot}0$ on the right.
3. Follow resistance circle through point 1·0 upward.
4. Locate point where it crosses reactance circle j2·0. Point of intersection is $z = 1{\cdot}0+j2{\cdot}0\,\Omega$.

Voltage standing wave ratio

Let the load impedance be $100-j50\,\Omega$ on a $50\,\Omega$ line. To find the VSWR.

1. Normalising Z_R yields $z_R = (100-j50)/50 = 2-j1{\cdot}0\,\Omega$.
2. Plot it on the chart.
3. Draw a circle with centre at point (1, 0) through point z_R.

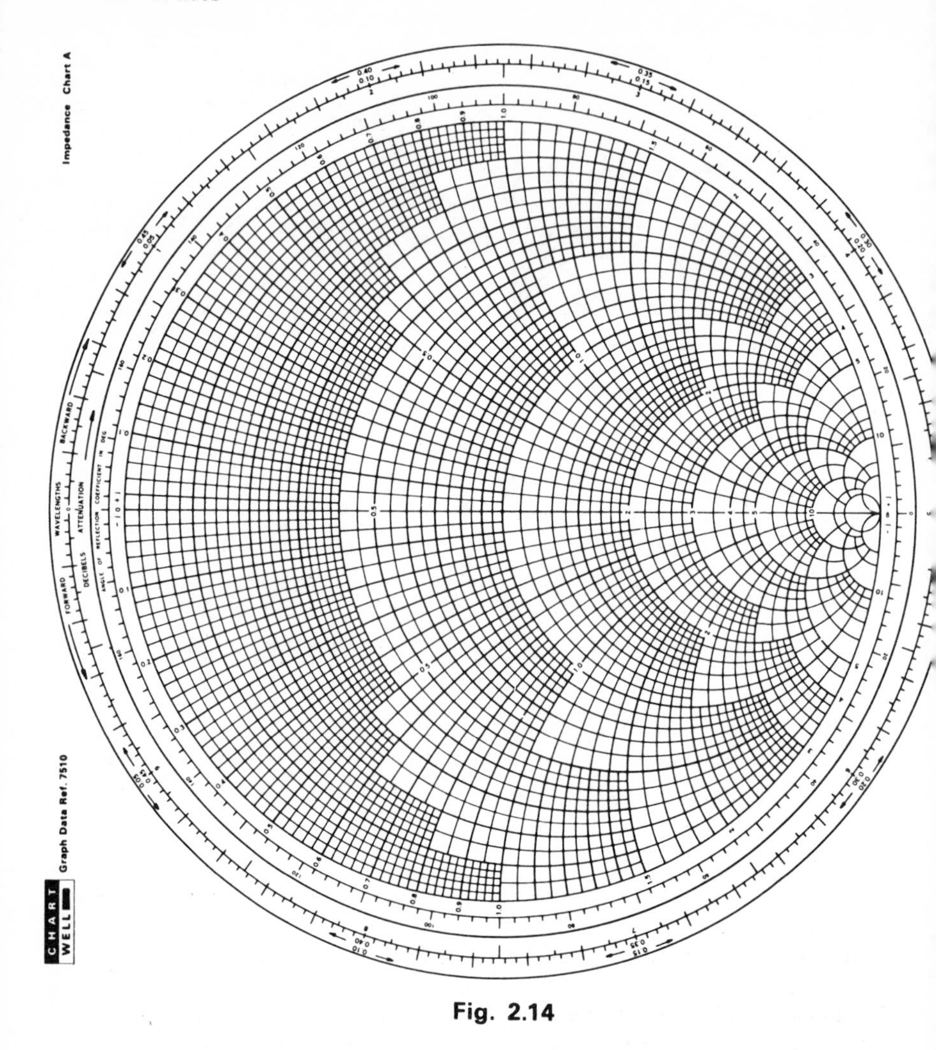

Fig. 2.14

4. Read point of intersection of the circle with the horizontal diameter on the right of the chart centre. This gives the VSWR as 2·6.

Reflection coefficient

Let the load impedance be 100 + j75 Ω on a 50 Ω line. To find the reflection coefficient.

1. Normalising Z_R yields $z_R = (100 + j75)/50 = 2 + j1{\cdot}5\ \Omega$.
2. Plot z_R.
3. Draw the VSWR circle through z_R and read the VSWR as 3·3.
4. From $|\rho| = (s - 1)/(s + 1)$, $|\rho|$ is obtained as 0·535.

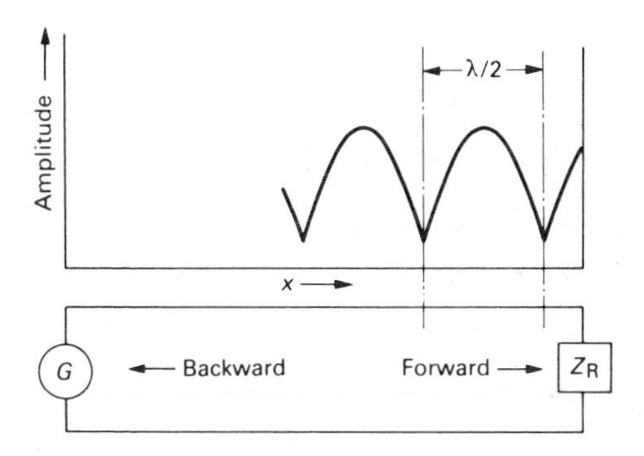

Fig. 2.15

5. Alternatively, scale-off the radius of the VSWR circle through point '3·3' and the unit circle radius also. The ratio of the smaller to larger radius is $|\rho| = 0{\cdot}535$.
6. Draw a radial line from chart centre through z_R to meet the phase angle circle. Read ϕ as 30°. Hence $\rho = 0{\cdot}535\underline{/30^\circ}$.

Admittance of a load

Let the load be $Z_R = 150 - j75\ \Omega$ on a $50\ \Omega$ line. To find Y_R.

1. Normalising Z_R yields $Z_R = (150 - j75)/50 = 3 - j1{\cdot}5\ \Omega$.
2. Plot z_R.
3. Draw the VSWR circle through z_R.
4. From point z_R draw a diameter through the chart centre to intersect the circle again.
5. Point of intersection is $y_R = 0{\cdot}27 + j0{\cdot}14$ which is the required normalised admittance.
6. Admittance $Y_R = y_R/50 = 0{\cdot}0054 + j0{\cdot}0028$ S.

Distance from load to voltage minimum

Let the load be $Z_R = 50 + j150\ \Omega$ on a $50\ \Omega$ line. To find the distance from Z_R to the first voltage minimum.

1. Normalising Z_R yields $z_R = (50 + j150)/50 = 1 + j3{\cdot}0\ \Omega$.
2. Plot z_R.
3. Draw radial line from chart centre through z_R to intersect wavelengths circle.
4. Locate point of voltage minimum as point $r = 0$ on the left of the chart. Note position on wavelengths circle as zero λ.
5. Move along wavelengths circle 'forwards'.
6. Read wavelength at point of intersection with radial line through z_R as $0{\cdot}296\,\lambda$. Hence, the first voltage minimum is $0{\cdot}296\,\lambda$ from the load.

Load impedance from shift of minima

To find the load impedance when a voltage minimum shifts $0{\cdot}1\lambda$ toward the load when the load is shorted. Let the VSWR on the $50\,\Omega$ line be 2·0.

1. When the load is shorted, first minimum occurs at the load. Since it moves $0{\cdot}1\lambda$ towards the load when the load is shorted, it must have been $0{\cdot}1\lambda$ away from the load.
2. Locate the short circuit $r = 0$ at the left hand end on the central chart diameter.
3. Move $0{\cdot}1\lambda$ towards load 'forwards', along wavelengths circle and draw radial line to chart centre.
4. Draw a VSWR circle of 2·0. Point of intersection with radial line is the load impedance $z_R = 0{\cdot}68 - j0{\cdot}48\,\Omega$.
5. Impedance $Z_R = 50(0{\cdot}68 - j0{\cdot}48) = 34{\cdot}0 - j24\,\Omega$.

3

Electromagnetic waves

3.1 Electromagnetic fields[16–18]

Electromagnetic waves may be transmitted along some form of guided structure and are called guided waves. Alternatively, they may be propagated through free space and are therefore unguided waves. In both cases, energy is carried by the electric and magnetic fields associated with the wave, hence the name electromagnetic wave.

However, at lower power frequencies it is more usual to talk about voltages and currents along a transmission line rather than the fields of the wave. Nevertheless, the energy is associated with the fields and the transmission line merely serves to guide the energy along. Such is the case for the two-wire line and coaxial cable, and the form of propagation is called a Transverse Electric and Magnetic wave or TEM wave as briefly outlined in the first chapter.

In this type of propagation the presence of the field is depicted by field lines whose direction is indicated by arrows and its intensity by the density of lines drawn. Typical field configurations for a two-wire line and a coaxial cable are shown in Fig. 3.1.

At very high frequencies, around 1 GHz, electromagnetic waves may under certain circumstances be transmitted along hollow metal guides

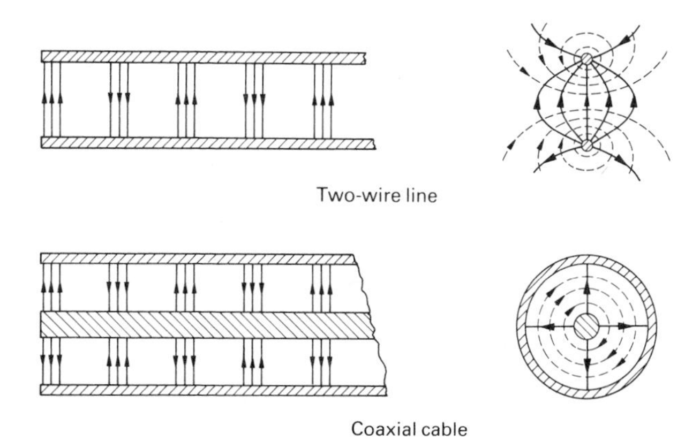

Fig. 3.1

called waveguides. Such guided waves are usually associated with fields rather than voltages and currents and the form of field propagation, though guided, is somewhat different from that of a TEM wave. It is called either a transverse electric (TE or *H* wave) or transverse magnetic (TM or *E* wave) as illustrated in Fig. 3.2.

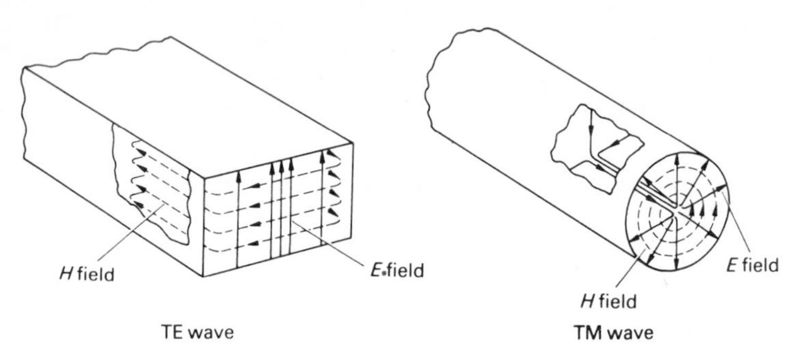

Fig. 3.2

Nevertheless, the basis of all such field phenomena is described in terms of certain vector quantities ***E***, ***D***, ***B***, ***H***, ***J*** which are formulated in a general form of field theory. The basis of such field theory may be purely electric such as the well known theory of Electrostatics or it may be purely magnetic and is known as Magnetostatics. A combination of both electric and magnetic effects due to moving charges or currents is called Electromagnetic Field Theory. Details of important concepts of Electrostatic and Magnetostatic theories are given in Ramo[16]. The development of electromagnetic field theory will now be considered.

3.2 Electromagnetic field theory

It is based on the five field vectors ***E***, ***D***, ***B***, ***H***, ***J***, whose basic equations are given by Ramo[16]. It is usual to refer to these quantities as vectors since they have both magnitude and direction. The essential study concerning the behaviour of vectors is known as vector analysis and is detailed in Ramo[16]. The basic ideas of electrostatic theory and magnetostatic theory were taken further and combined into a single unified theory of electromagnetic waves by Maxwell and are concisely expressed in his well known equations.

(a) Maxwell's equations

These are four basic equations which apply to time-varying (alternating) fields and electromagnetic phenomena. The theory is a generalisation of the previous work of Faraday, Ampere and Gauss. Maxwell was able to build upon their ideas and to make a further contribution himself which led him to formulating a coherent theory for the field vectors $\boldsymbol{E}$, $\boldsymbol{D}$, $\boldsymbol{B}$, $\boldsymbol{H}$ and $\boldsymbol{J}$. Further details are given in Stratton[17].

In the case of alternating fields $\boldsymbol{E}$ and $\boldsymbol{H}$, we can summarise Maxwell's equations for a conducting medium and a dielectric medium in the following manner.

Conducting medium

$$\operatorname{curl} \boldsymbol{H} = \boldsymbol{J} + \frac{\partial \boldsymbol{D}}{\partial t} = (\sigma + \mathrm{j}\omega\varepsilon)\boldsymbol{E}$$

$$\operatorname{curl} \boldsymbol{E} = -\frac{\partial \boldsymbol{B}}{\partial t} = -\mathrm{j}\omega\mu\boldsymbol{H}$$

$$\operatorname{div} \boldsymbol{D} = \rho$$

$$\operatorname{div} \boldsymbol{B} = 0$$

Dielectric medium

$$\operatorname{curl} \boldsymbol{H} = \frac{\partial \boldsymbol{D}}{\partial t} = \mathrm{j}\omega\varepsilon\boldsymbol{E}$$

$$\operatorname{curl} \boldsymbol{E} = \frac{\partial \boldsymbol{B}}{\partial t} = -\mathrm{j}\omega\mu\boldsymbol{H}$$

$$\operatorname{div} \boldsymbol{D} = 0$$

$$\operatorname{div} \boldsymbol{B} = 0$$

since σ and ρ are both zero in a dielectric medium.

(b) The wave equation

For a dielectric medium we have

$$\operatorname{curl} \boldsymbol{E} = \nabla \times \boldsymbol{E} = -\mathrm{j}\omega\mu\boldsymbol{H}$$

Hence $$\nabla \times \nabla \times \boldsymbol{E} = \operatorname{curl}(-\mathrm{j}\omega\mu\boldsymbol{H}) = -\mathrm{j}\omega\mu \operatorname{curl} \boldsymbol{H}$$

or $$\nabla \times \nabla \times \boldsymbol{E} = \omega^2\mu\varepsilon\boldsymbol{E}$$

Also $$\nabla(\nabla \cdot \boldsymbol{E}) - \nabla^2\boldsymbol{E} = \nabla \times \nabla \times \boldsymbol{E}$$

with $$\nabla \cdot \boldsymbol{D} = \nabla \cdot \varepsilon\boldsymbol{E} = 0$$

Hence $$\nabla(0) - \nabla^2\boldsymbol{E} = \omega^2\mu\varepsilon\boldsymbol{E}$$

or $$\nabla^2\boldsymbol{E} = -\omega^2\mu\varepsilon\boldsymbol{E}$$

Now $1/\sqrt{\mu\varepsilon}$ has the dimensions of a velocity from which Maxwell concluded that light is an electromagnetic wave propagated through free space with velocity $c = 1/\sqrt{\mu\varepsilon}$.

Hence $$\nabla^2\boldsymbol{E} = (-\omega^2/c^2)\boldsymbol{E}$$

For alternating fields we have also

$$\boldsymbol{E} = \boldsymbol{E}_0 \sin \omega t$$

or
$$\frac{\partial^2 \boldsymbol{E}}{\partial t^2} = -\omega^2 \boldsymbol{E}_0 \sin \omega t = -\omega^2 \boldsymbol{E}$$

Hence
$$\nabla^2 \boldsymbol{E} = 1/c^2 \frac{\partial^2 \boldsymbol{E}}{\partial t^2}$$

Similarly
$$\nabla^2 \boldsymbol{H} = \frac{1}{c^2} \frac{\partial^2 \boldsymbol{H}}{\partial t^2}$$

which are the wave equations for the alternating field vectors $\boldsymbol{E}$ and $\boldsymbol{H}$.

In a *conducting* medium without free electrical charges, we have finite σ with $\rho = 0$.

Hence
$$\nabla \times \boldsymbol{E} = -\mathrm{j}\omega\mu \boldsymbol{H}$$

or
$$\nabla \times \nabla \times \boldsymbol{E} = -\mathrm{j}\omega\mu \operatorname{curl} \boldsymbol{H}$$
$$= -\mathrm{j}\omega\mu(\sigma + \mathrm{j}\omega\varepsilon)\boldsymbol{E}$$

Since
$$\nabla \times \nabla \times \boldsymbol{E} = \nabla(\nabla \cdot \boldsymbol{E}) - \nabla^2 \boldsymbol{E}$$

with
$$\nabla \cdot \boldsymbol{D} = \nabla \cdot \varepsilon \boldsymbol{E} = 0$$

Hence
$$\nabla^2 \boldsymbol{E} = -\nabla \times \nabla \times \boldsymbol{E} = (\mathrm{j}\omega\mu\sigma - \omega^2\mu\varepsilon)\boldsymbol{E}$$

or
$$\nabla^2 \boldsymbol{E} = \gamma^2 \boldsymbol{E}$$

where
$$\gamma^2 = (\mathrm{j}\beta)^2 = -\beta^2 = (\mathrm{j}\omega\mu\sigma - \omega^2\mu\varepsilon)$$

or
$$\beta = \sqrt{\omega^2\mu\varepsilon - \mathrm{j}\omega\mu\sigma}$$

Similarly
$$\nabla^2 \boldsymbol{H} = \gamma^2 \boldsymbol{H}$$

Also since
$$\boldsymbol{E} = \boldsymbol{E}_0 \sin \omega t$$

$$\frac{\partial \boldsymbol{E}}{\partial t} = \omega \boldsymbol{E}_0 \cos \omega t = \mathrm{j}\omega \boldsymbol{E}$$

$$\frac{\partial^2 \boldsymbol{E}}{\partial t^2} = -\omega^2 \boldsymbol{E}_0 \sin \omega t = -\omega^2 \boldsymbol{E}$$

with
$$\omega^2\mu\varepsilon = \omega^2/c^2 = -1/c^2 \frac{\partial^2}{\partial t^2}$$

Hence
$$\nabla^2 \boldsymbol{E} = \gamma^2 \boldsymbol{E} = \sigma\mu \frac{\partial \boldsymbol{E}}{\partial t} + \frac{1}{c^2} \frac{\partial^2 \boldsymbol{E}}{\partial t^2}$$

$$\nabla^2 \boldsymbol{H} = \gamma^2 \boldsymbol{H} = \sigma\mu \frac{\partial \boldsymbol{H}}{\partial t} + \frac{1}{c^2} \frac{\partial^2 \boldsymbol{H}}{\partial t^2}$$

(c) Power flow

The energy stored per unit volume in an electrostatic field is $\frac{1}{2}\varepsilon E^2$ where $\boldsymbol{E}$ is the electric field vector. For a magnetic field it is $\frac{1}{2}\mu H^2$ where $\boldsymbol{H}$ is the magnetic field vector. For an electromagnetic field, the most general expression is the sum of the electric and magnetic energies.

Hence $$\left.\begin{array}{c}\text{Total energy stored}\\ \text{per unit volume}\end{array}\right\} = \tfrac{1}{2}(\varepsilon E^2 + \mu H^2)/\text{m}^3$$

Consider now, the surface d$\boldsymbol{a}$ shown in Fig. 3.3 with outward flow $\mathscr{P}$ per unit area.

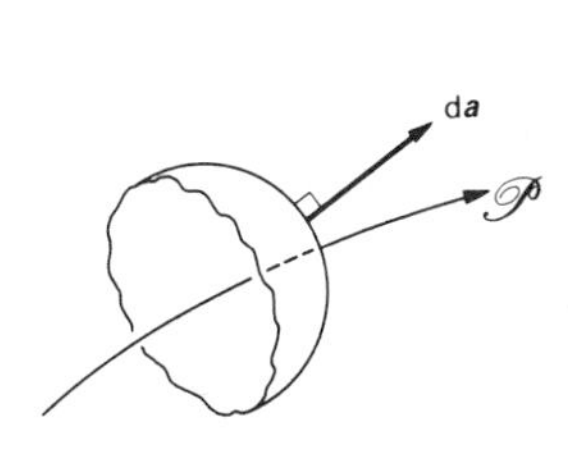

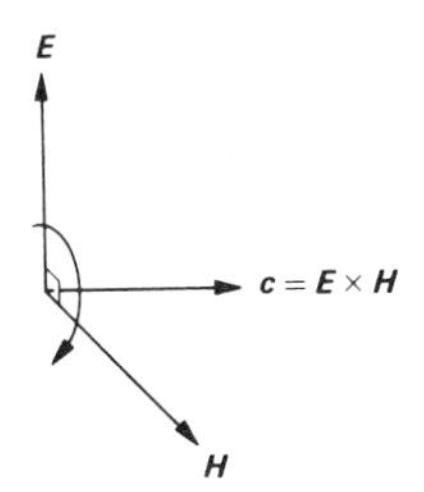

Fig. 3.3

Total power flow through d$\boldsymbol{a}$ = $\oint_s \mathscr{P} \cdot \mathrm{d}\boldsymbol{a}$. This leads to a loss of stored energy given by

$$\text{loss} = \frac{-\partial}{\partial t}\left[\frac{1}{2}\oint_V (\varepsilon E^2 + \mu H^2)\,\mathrm{d}v\right] \qquad \text{(magnitude only)}$$

or $$\oint_S \mathscr{P}\cdot\mathrm{d}\boldsymbol{a} = -\oint_V \left(\mu H\frac{\partial \boldsymbol{H}}{\partial t} + \varepsilon E\frac{\partial \boldsymbol{E}}{\partial t}\right)\mathrm{d}v$$

$$= -\oint_V \left(\mu \boldsymbol{H}\cdot\frac{\partial \boldsymbol{H}}{\partial t} + \varepsilon \boldsymbol{E}\cdot\frac{\partial \boldsymbol{E}}{\partial t}\right)\mathrm{d}v$$

But $$\nabla\times\boldsymbol{E} = -\mu\frac{\partial \boldsymbol{H}}{\partial t} \quad\text{and}\quad \nabla\times\boldsymbol{H} = \varepsilon\frac{\partial \boldsymbol{E}}{\partial t}$$

for a dielectric medium.

Hence $$\oint_S \mathscr{P}\cdot\mathrm{d}\boldsymbol{a} = \oint_V [\boldsymbol{H}\cdot(\nabla\times\boldsymbol{E}) - \boldsymbol{E}\cdot(\nabla\times\boldsymbol{H})]\,\mathrm{d}v$$

Also $$\nabla\cdot(\boldsymbol{E}\times\boldsymbol{H}) = \boldsymbol{H}\cdot(\nabla\times\boldsymbol{E}) - \boldsymbol{E}\cdot(\nabla\times\boldsymbol{H})$$

Hence $$\oint_S \mathscr{P}\cdot\mathrm{d}\boldsymbol{a} = \oint_V \nabla\cdot(\boldsymbol{E}\times\boldsymbol{H})\,\mathrm{d}v = \oint_S (\boldsymbol{E}\times\boldsymbol{H})\,\mathrm{d}\boldsymbol{a}$$

or $$\mathscr{P} = \boldsymbol{E}\times\boldsymbol{H}\ \text{watts/m}^2$$

which is Poynting's theorem.

For alternating quantities we obtain

$$\mathscr{P}_{av} = \tfrac{1}{2}(\boldsymbol{E} \times \boldsymbol{H}^*) \text{ watts/m}^2$$

where $\boldsymbol{H}^*$ is the conjugate of $\boldsymbol{H}$. This gives the *real power* propagated with velocity c as shown in Fig. 3.3.

Example 3.1

A plane TEM wave is travelling through free space in the x direction with velocity c. Assuming sinusoidal variations of the electric and magnetic fields, deduce the intrinsic impedance of free space.

Solution

Assuming rectangular coordinates, for the $\boldsymbol{E}$ vector we have

$$\boldsymbol{E} = \boldsymbol{i}E_x + \boldsymbol{j}E_y + \boldsymbol{k}E_z$$

with

$$\nabla^2 \boldsymbol{E} = \frac{1}{c^2}\frac{\partial^2 \boldsymbol{E}}{\partial t^2}$$

Hence

$$\frac{\partial^2 E_x}{\partial x^2} + \frac{\partial^2 E_x}{\partial y^2} + \frac{\partial^2 E_x}{\partial z^2} = \frac{1}{c^2}\frac{\partial^2 E_x}{\partial t^2}$$

$$\frac{\partial^2 E_y}{\partial x^2} + \frac{\partial^2 E_y}{\partial y^2} + \frac{\partial^2 E_y}{\partial z^2} = \frac{1}{c^2}\frac{\partial^2 E_y}{\partial t^2}$$

$$\frac{\partial^2 E_z}{\partial x^2} + \frac{\partial^2 E_z}{\partial y^2} + \frac{\partial^2 E_z}{\partial z^2} = \frac{1}{c^2}\frac{\partial^2 E_z}{\partial t^2}$$

Since the plane wave is travelling along the x direction, it has only components E_y and H_z as shown in Fig. 3.4.

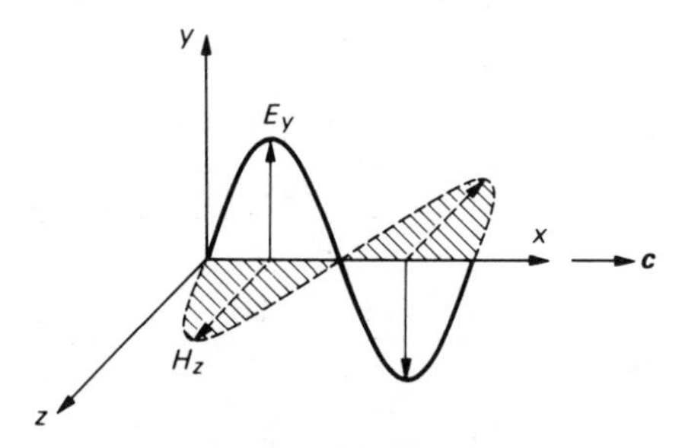

Fig. 3.4

Hence

$$E_x = E_z = 0$$

$$\frac{\partial}{\partial y} = \frac{\partial}{\partial z} = 0$$

with

$$\frac{\partial^2 E_y}{\partial x^2} = \frac{1}{c^2}\frac{\partial^2 E_y}{\partial t^2}$$

The last equation is similar to that for a transmission line. The general solution is of the form

$$E_y = (Ae^{-\gamma x} + Be^{\gamma x})e^{j\omega t}$$

As there is no reflected wave in free space, we obtain

$$E_y = Ae^{-\gamma x}e^{j\omega t}$$

If the peak value of the field is E_0 at $x = 0$, then

$$E_y = E_0 e^{-\gamma x}e^{j\omega t}$$

For free space $\alpha \simeq 0$ and so $\gamma = j\beta$ with $\beta = 2\pi/\lambda$ where λ is the free space wavelength.

Hence $$E_y = E_0 e^{-j\beta x}e^{j\omega t} = E_0 e^{j(\omega t - \beta x)}$$

Similarly, starting from the equation

$$\frac{\partial^2 H_z}{\partial x^2} = \frac{1}{c^2}\frac{\partial^2 H_z}{\partial t^2}$$

we obtain $H_z = H_0 e^{j(\omega t - \beta x)}$

H_z component

Since curl $\boldsymbol{H} = \varepsilon_0(\partial \boldsymbol{E}/\partial t)$ for free space, we have

$$-j\frac{\partial H_z}{\partial x} = j\varepsilon_0(\partial/\partial t)[E_0 e^{j(\omega t - \beta x)}]$$

$$= j\varepsilon_0 j\omega E_0 e^{j(\omega t - \beta x)}$$

$$= j\varepsilon_0 j\omega E_y$$

or $$\frac{\partial H_z}{\partial x} = -\varepsilon_0 j\omega E_y$$

so $$-j\beta H_0 e^{-j\beta x}e^{j\omega t} = -j\beta H_z = -\varepsilon_0 j\omega E_y$$

with $$H_z = \frac{-\varepsilon_0 j\omega E_y}{-j\beta} = \varepsilon_0(\omega/\beta)E_y = \varepsilon_0 c E_y$$

Hence $$\frac{E_y}{H_z} = \frac{E_y}{\varepsilon_0 c E_y} = \frac{1}{\varepsilon_0 c} = \sqrt{\frac{\mu_0}{\varepsilon_0}} = 377\ \Omega$$

or $$Z_0 = 377\ \Omega$$

where Z_0 is called the intrinsic impedance of free space.

3.3 Boundary conditions

For time-varying fields, certain boundary conditions exist at a conducting surface or at a dielectric interface which must satisfy Maxwell's equations.

Dielectric interface

If E_{t_1}, E_{t_2} and H_{t_1}, H_{t_2} are the tangential electric and magnetic components respectively, at the interface of the two media shown in Fig. 3.5, we must have

$$E_{t_1} = E_{t_2}$$

$$H_{t_1} = H_{t_2}$$

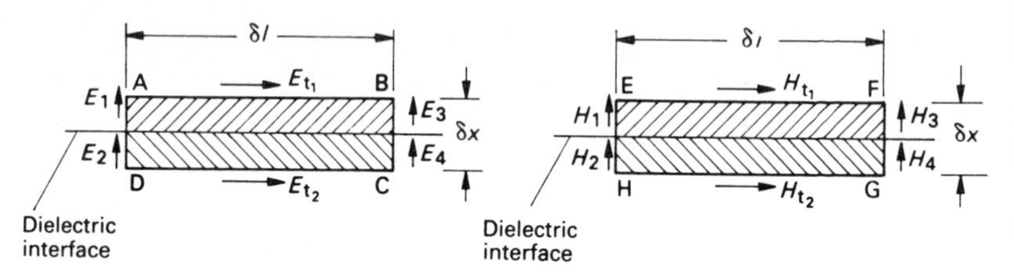

Fig. 3.5

Similarly, if D_{n_1}, D_{n_2} and B_{n_1}, B_{n_2} are the normal components of electric and magnetic flux density respectively, we must have

$$D_{n_1} = D_{n_2}$$

$$B_{n_1} = B_{n_2}$$

The proofs of these expressions are given in Ramo[16].

Conducting surface

At the surface of a perfect conductor for which $\sigma = 0$, we have $E_{t_2} = 0$ since no electric field exists in an *equipotential* surface. Hence

$$E_{t_1} = E_{t_2} = 0$$

and so no tangential components exist near a perfect conductor. Normal components can exist as shown in Fig. 3.6. Furthermore, as there are no magnetic poles at the surface we have $B_{n_2} = 0$. Hence

$$B_{n_1} = B_{n_2} = 0$$

and so the magnetic lines form closed loops as shown in Fig. 3.6.

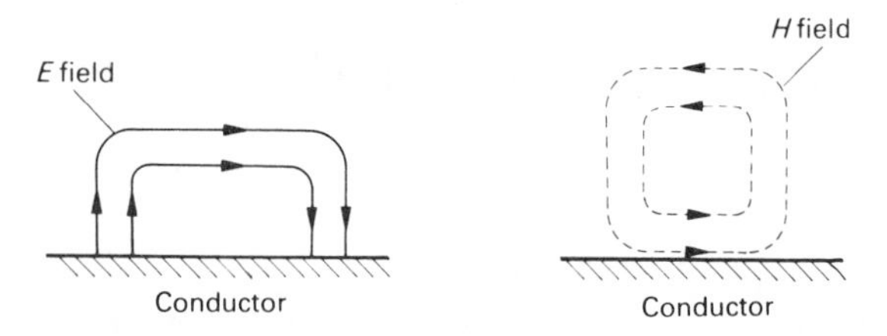

Fig. 3.6

Comment
The boundary conditions are useful in determining the field patterns inside bounded regions such as waveguides.

Surface penetration

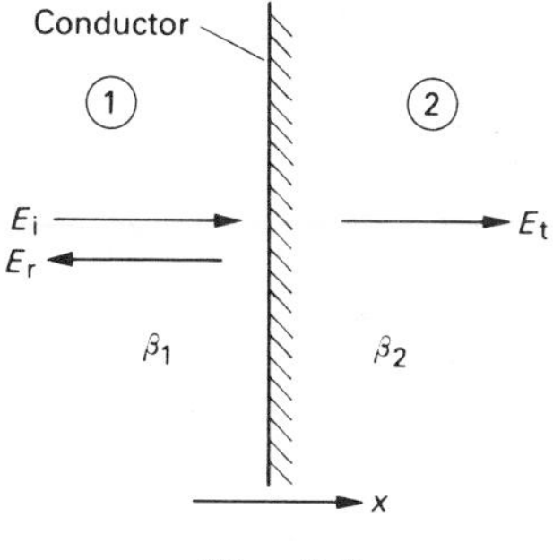

Fig. 3.7

Consider a plane wave incident normally at the conductor boundary in Fig. 3.7. Let the incident wave be denoted by

$$E_i = E_1\, e^{j(\omega t - \beta_1 x)}$$

and the transmitted wave by

$$E_t = E_2 e^{j(\omega t - \beta_2 x)}$$

where β_1 and β_2 are the phase-change coefficients outside and inside the conductor respectively. The reflected wave E_r is then given by

$$E_r = E_i - E_t$$

In general, the phase-change coefficient for a conducting medium is given by

$$\beta = \sqrt{\omega^2 \mu \varepsilon - j\sigma\omega\mu}$$

where σ is the conductivity.
For metals

$$j\sigma\omega\mu \gg \omega^2 \mu \varepsilon$$

Hence
$$\beta_2 \simeq \sqrt{-j\sigma\omega\mu}$$

$$\simeq \sqrt{\frac{\sigma\omega\mu}{2}}\sqrt{-2j}$$

$$\simeq \sqrt{\frac{\sigma\omega\mu}{2}}(1 - j)$$

or $$\beta_2 \simeq s(1-\mathrm{j})$$

where $$s = \sqrt{\frac{\sigma\omega\mu}{2}}$$

Hence $$E_t = E_2 e^{(\mathrm{j}\omega t - \mathrm{j}sx - sx)} = E_2 e^{-sx} e^{\mathrm{j}(\omega t - sx)}$$

The penetration or skin depth δ is defined such that the amplitude E_t in the conductor is reduced to 1/e of its initial value E_2. This happens when $sx = 1$ with $x = \delta$. Hence

$$s\delta = 1$$

or $$\delta = \sqrt{\frac{2}{\sigma\omega\mu}}$$

3.4 Reflected and refracted waves

When an electromagnetic wave is incident on a dielectric boundary, like an air-glass surface, and at an oblique angle, both reflection and refraction take place as in the case of light waves. Thus, much of the consideration of light waves can be applied to a general electromagnetic wave provided the wavelength of the wave is small in comparison with the physical dimensions of the boundary surface. In this case, it is convenient to use geometrical optics which employs simple ray theory by means of straight lines rather than by *E* and *H* fields.

Consider for example an electromagnetic wave incident at a point O on a dielectric surface at an oblique angle *i* as shown in Fig. 3.8. The incident wave is AO at an angle of incidence *i* with the normal, the reflected wave is OB at an angle of reflection *i′* with the normal and the refracted wave is OC

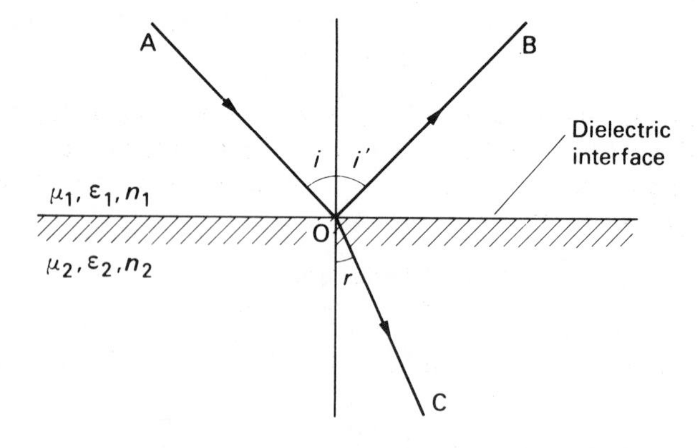

Fig. 3.8

at an angle r with the normal. Let the constants of the two media be denoted by $\mu_1, \varepsilon_1, n_1$ and $\mu_2, \varepsilon_2, n_2$ where μ_1, μ_2 are the permeabilities, $\varepsilon_1, \varepsilon_2$ are the permittivities and n_1, n_2 are the refractive indices of the two media.

In the case of pure reflection, if the incident wave is a plane wave and the dielectric boundary is a smooth, plane surface, the laws of reflection state that the incident wave, the reflected wave and the normal, all lie in the same plane. Furthermore, the angle of incidence i is equal to the angle of reflection i'. However, the incident energy in general may not be equal to the reflected energy and the ratio of the reflected energy to the incident energy is expressed by the *reflection coefficient* $\rho e^{j\phi}$ where $0 < \rho < 1$ and ϕ is the phase shift produced at the point of reflection O.

A practical case of interest is ground reflection, where the reflecting boundary is the earth's surface. For a smooth, flat earth, there is only a *specular* component on reflection and the reflection coefficient for grazing incidence is close to unity. For a rough surface however, it is much less than 1. This is because of a *diffuse* component which is due to energy being scattered in various directions. Furthermore, the behaviour of the phase angle can vary, depending on the angle of incidence and the polarization of the wave.

In the case of *horizontal* polarisation, ϕ is nearly always 180°, but for *vertical* polarisation, this value of ϕ is true only at low grazing incidence. Reflections from the earth's surface are of practical importance in radar systems and give rise to such effects as multipath interference and the fading of signals.

When the dielectric boundary is an air-glass interface, refraction also takes place and if $n_1 > n_2$, the angle of incidence i is greater than the angle of refraction r i.e. the refracted ray is bent *towards* the normal. The relationship between the angle of incidence i and the angle of refraction r is given by Snell's law which states simply that

$$\frac{\sin i}{\sin r} = \frac{v_1}{v_2}$$

where v_1 is the velocity of the wave in the first medium and v_2 is the velocity of the wave in the second medium. For an air-glass interface, $v_1 = c$ and by definition the refractive index n of a medium such as glass is given by $n = c/v_2$. Thus, we obtain

$$\frac{\sin i}{\sin r} = \frac{c}{v_2} = n$$

and as n is related to the dielectric constants ε_1, ε_2 by the relationship

$$n = \sqrt{\frac{\mu_2 \varepsilon_2}{\mu_1 \varepsilon_1}} = \sqrt{\frac{\varepsilon_2}{\varepsilon_1}}$$

since $\mu_1 \simeq \mu_2$ for an air-glass interface, we have

$$\frac{\sin i}{\sin r} = n = \sqrt{\frac{\varepsilon_2}{\varepsilon_1}}$$

Two important cases of interest will now be considered. In the first case, the incident angle i is greater than the *critical* angle and no refraction takes place but only pure reflection. The condition for total internal reflection occurs when

$$i_c = \sin^{-1}\sqrt{\frac{\varepsilon_2}{\varepsilon_1}}$$

A practical application of this phenomenon is in the optical fibre where light waves are reflected from the boundary surface and propagate down the glass fibre with no loss at the point of reflection, as no energy exists outside the fibre core.

The second case of interest occurs when the total angle $(i+r) = 90°$ i.e. $r = (90-i)$ and we obtain

$$\frac{\sin i}{\sin r} = \frac{\sin i}{\cos i} = \tan i = \sqrt{\frac{\varepsilon_2}{\varepsilon_1}}$$

or

$$i = \tan^{-1}\sqrt{\frac{\varepsilon_2}{\varepsilon_1}}$$

which is known as the *Brewster angle*[19]. In this case, unpolarised light emerges as vertically polarised light, a phenomenon which finds application in the glass-air windows used in optical lasers. For an air-glass interface the Brewster angle $i \simeq 56°$ and $r = 34°$.

The phenomenon of refraction also has important significance in radar applications. Due to the variation in the refractive index of the atmosphere, it causes electromagnetic waves to be bent downwards towards the earth, thus increasing radar range at low altitudes. However, it gives rise to errors in the altitude of a distant target. Furthermore, as the refractive index of the ionosphere is less than unity, electromagnetic waves incident on the ionosphere are also bent back to earth, thus making long distance communication possible.

Example 3.2

An electromagnetic wave at a frequency of 2 GHz is incident vertically on a sheet of copper with a conductivity of $5{\cdot}8 \times 10^7$ S/m. If the field strength of the vertically polarised wave is 1 V/m, determine

(a) the incident power density
(b) the depth of penetration
(c) the wave impedance
(d) the wave velocity in the metal sheet.

Solution

Let the incident wave be denoted by E_i and the transmitted wave by E_t as is shown in Fig. 3.7. Also, let the permeability and conductivity in the metal be denoted by μ and σ respectively.

(a) The incident power density P is given by

$$P = E \times H \text{ watts/m}^2 = E_i^2/377 \text{ watts/m}^2$$

or

$$P = 1/377 = 0{\cdot}002\,65 \text{ watts/m}^2$$

(b) The depth of penetration in the copper sheet is given by

$$\delta = \sqrt{\frac{2}{\sigma\omega\mu}} = \sqrt{\frac{2}{5{\cdot}8 \times 10^{-7} \times 2\pi \times 2 \times 10^9 \times 4\pi \times 10^{-7}}}$$

or

$$\delta = 1{\cdot}476 \times 10^{-6} \text{ metre}$$

(c) The wave impedance in copper is given by

$$Z = \frac{\omega\mu}{\beta}$$

where

$$\beta \simeq \sqrt{-j\sigma\omega\mu} \simeq \sqrt{\frac{\sigma\omega\mu}{2}}\,\underline{/-45^\circ}$$

Thus

$$Z = \frac{\omega\mu}{\sqrt{\frac{\sigma\omega\mu}{2}}\,\underline{/-45^\circ}} = \sqrt{\frac{\omega\mu}{\sigma}}\,\underline{/45^\circ}$$

or

$$Z = \sqrt{\frac{2\pi \times 2 \times 10^9 \times 4\pi \times 10^{-7}}{5{\cdot}8 \times 10^7}}\,\underline{/45^\circ} = 0{\cdot}0165\underline{/45^\circ} \text{ ohm}$$

(d) The wave velocity in the metal is given by

$$v_p = \frac{\omega}{\beta_{real}}$$

where

$$\beta_{real} = \sqrt{\frac{\sigma\omega\mu}{2}}$$

Thus

$$v_p = \frac{\omega}{\sqrt{\sigma\omega\mu/2}} = \sqrt{\frac{2\omega}{\sigma\mu}}$$

$$= \sqrt{\frac{2 \times 2\pi \times 2 \times 10^9}{5{\cdot}8 \times 10^7 \times 4\pi \times 10^{-7}}}$$

or

$$v_p = 18\,570 \text{ m/s}$$

4

Waveguide theory

Electromagnetic waves may be transmitted along hollow tubes or waveguides under certain conditions. Such guided waves are unlike those transmitted by two-wire lines or coaxial cables and are of main interest at microwave frequencies.

The waveguide is essentially a hollow metal tube either of rectangular or circular cross section as shown in Fig. 3.2 and is made of copper, brass or aluminium. For special applications, other shapes of guide may also be used, e.g. square waveguides. The mode of propagation is essentially in terms of the electric and magnetic field as there is no centre conductor, while the hollow guide merely serves to guide the electromagnetic wave. Occasionally, reference may be made to the voltage across the guide or to the induced currents in the walls of the guide as they throw some light on power transmitted by the waveguide or on the power losses in the walls of the waveguide.

4.1 Waveguide transmission[20,21]

It is characterised by two important features.

(a) There is a minimum frequency below which a given wavelength will not transmit the wave. It is called the cut-off frequency f_c and is directly related to the size of waveguide used.

(b) There is always a component of $\boldsymbol{E}$ or $\boldsymbol{H}$ along the direction of propagation.

The guided waves may be propagated along the waveguide with different field patterns called 'modes'. Modes are mainly of two types, those which have an $\boldsymbol{E}$ component along the direction of propagation are called TM or *E waves*, while those with an $\boldsymbol{H}$ component along the direction of propagation are called TE or *H waves*. To distinguish between the various types of modes, subscripts m and n are used as defined hereafter.

Rectangular modes

The two types TE_{mn} (H_{mn}) modes and TM_{mn} (E_{mn}) modes, where m, n are integers. The subscript m refers to the number of *half* sinusoidal variations of the field along dimension 'a', while the subscript n refers to the number of *half* sinusoidal variations of the field along dimension 'b', e.g. $TE_{10}(H_{10})$, $TM_{11}(E_{11})$.

Circular modes

The two types are $TE_{mn}(H_{mn})$ modes and $TM_{mn}(E_{mn})$ modes where m, n are integers. The subscript m refers to the number of *full* sinusoidal variations of the field along the circumference and the subscript n refers to the number of *half* sinusoidal variations of the field along a radius, e.g. $TE_{01}(H_{01})$, $TM_{12}(E_{12})$.

Basically, a rectangular waveguide may be considered as a form of parallel plate transmission line between which the wave is trapped. It is therefore possible to propagate a waveguide mode by the synthesis of two plane waves travelling towards each other at an angle θ to the horizontal provided the boundary conditions given in Section 3.3 are satisfied.

Consider, for example, the two plane waves shown in Fig. 4.1 which are travelling at an angle θ with the horizontal. For constructional reasons, magnetic field lines are considered, spaced $\lambda/2$ apart rather than electric field lines.

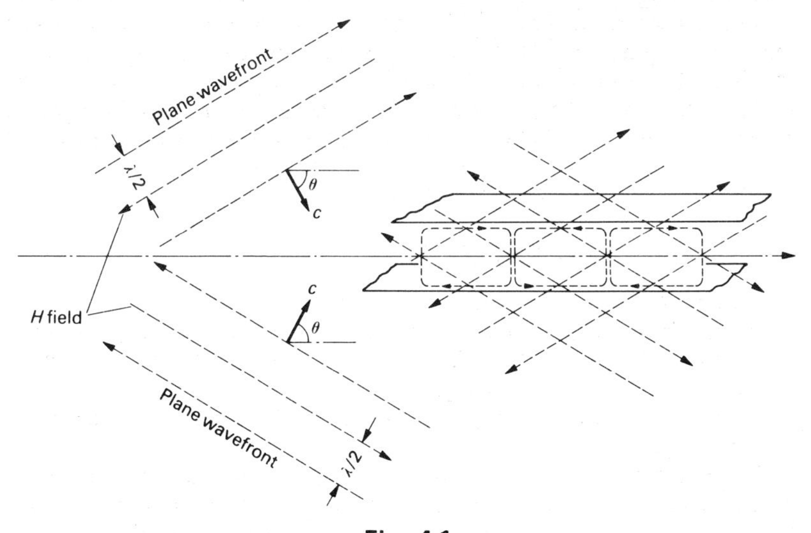

Fig. 4.1

When the two waves overlap as shown on the right, the resultant magnetic field lines at the intersecting points form closed loops along the axis of the diagram. By placing parallel plates tangential to the sides of a single set of such loops, a waveguide mode can be shown to exist within such a parallel plate structure. Side plates may then be added to form a closed waveguide with the wave travelling from left to right along the waveguide. The predominant $TE_{10}(H_{10})$ mode is shown in Fig. 4.4.

4.2 Phase and group velocities

The component plane waves travel with the velocity of light c towards one another, but the composite wave pattern formed when they overlap, travels along the guide with a different velocity called the phase or guide velocity v_p.

To illustrate this, let the wavefronts be represented by OM and ON as in Fig. 4.2 and after one second let them travel through distances MP and NP at angle θ to the axis. In the same time, the point of intersection O on the wavefronts has travelled to P with velocity v_p.

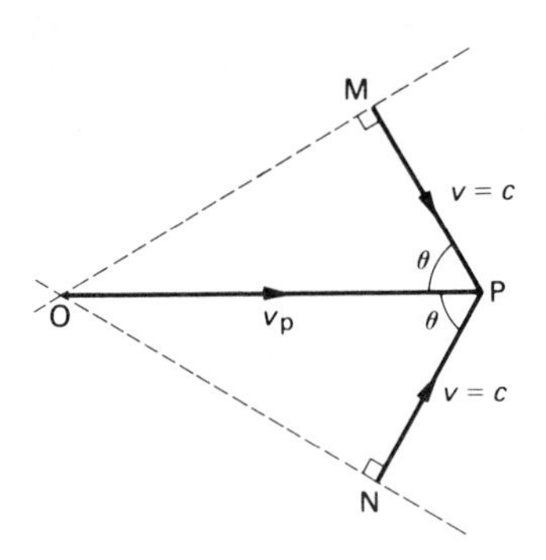

Fig. 4.2

From the diagram we have

$$v/v_p = \cos\theta$$

or

$$v_p = v/\cos\theta$$

As $\cos\theta < 1$, hence $v_p > c$

If λ is the free space wavelength and λ_g is the guide wavelength we have

$$v = f\lambda$$

$$v_p = f\lambda_g$$

or

$$v/v_p = \lambda/\lambda_g = \cos\theta \quad \text{from above}$$

As $\cos\theta < 1$, hence $\lambda_g > \lambda$

Since $\lambda_g > \lambda$, the composite waveguide pattern travels in the waveguide with a phase velocity $v_p > c$. However, this does not violate the theory of relativity since the energy of the wave is propagated along the waveguide with a group velocity v_g where v_g is the horizontal component of the free space velocity of each component plane wave.

Hence $$v_g = v\cos\theta$$

$$v_p v_g = (v/\cos\theta)v\cos\theta = v^2 = c^2$$

or $$v_p v_g = c^2$$

4.3 Waveguide equation

Consider the propagation of the H_{10} mode in a rectangular guide with dimensions a, b as shown in Fig. 4.3

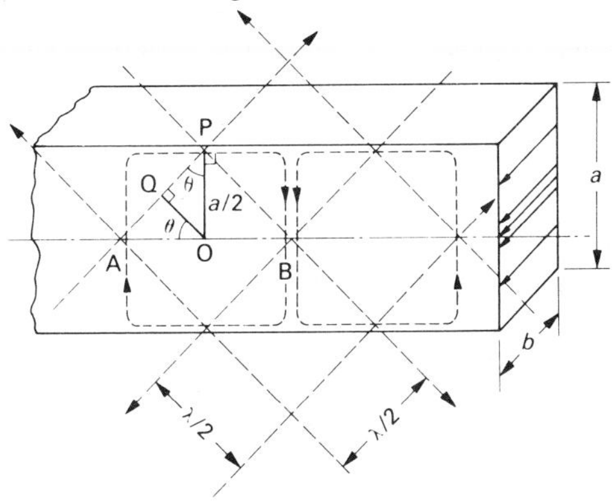

Fig. 4.3

From the diagram OP and OQ are perpendiculars from the centre point O such that

$$OP = a/2$$

$$OQ = \lambda/4$$

From △OQP we have

$$\sin\theta = OQ/OP = \frac{\lambda/4}{a/2} = \frac{\lambda}{2a}$$

But $$\cos\theta = v/v_p = (c/f)\lambda_g = \lambda/\lambda_g$$

and $$\sin^2\theta + \cos^2\theta = 1$$

Hence $$(\lambda/2a)^2 + (\lambda/\lambda_g)^2 = 1$$

Now, if λ_c is the cut-off wavelength, $\lambda_c = 2a$ for the $TE_{10}(H_{10})$ mode.*

Hence $$(\lambda/\lambda_c)^2 + (\lambda/\lambda_g)^2 = 1$$

or $$1/\lambda^2 = 1/\lambda_c^2 + 1/\lambda_g^2$$

* See Section 4.5.

where λ is the free space wavelength, λ_c is the cut-off wavelength and λ_g is the guide wavelength. This is the basic waveguide equation relating λ, λ_c and λ_g. It can be shown to be true for any mode, either in rectangular or circular waveguide.

Example 4.1

What factors influence the choice of the dimensions for a rectangular waveguide used to transmit an H_{10} wave? Explain the meaning of phase velocity and group velocity in relation to the field-pattern of the wave, and demonstrate from the geometry of the pattern, or otherwise, that the product of these velocities is the square of the velocity of light. Calculate the velocities for this wave at a frequency of 10 000 MHz in a guide which measures $2{\cdot}54\ \text{cm} \times 1{\cdot}27\ \text{cm}$ internally and is air filled. (U.L.)

Solution

The dimensions a and b are chosen for propagation of the H_{10} mode at the given frequency f. Since the free space wavelength $\lambda < \lambda_c$ where $\lambda_c = 2a$ for the H_{10} mode, hence $\lambda < 2a$. However, to avoid propagation of the next higher mode H_{20} where $\lambda_c = a$, we must have $\lambda > a$. Hence $\lambda > a > \lambda/2$, the exact value being chosen to allow for $\pm 20\%$ operation on either side of the operating frequency, with minimum attenuation.

Dimension b on the other hand, is not critical, but it determines the power handling capacity of the waveguide since the breakdown voltage across the top and bottom walls of the waveguide, depends upon b. However, b must not be too large or else it will propagate the H_{01} mode. Usually $b = a/2$ in practice, as this gives a value of guide impedance $Z_{TE} = 377\,\lambda_g/\lambda$.

The phase and group velocities v_p and v_g respectively have been explained in the text in Section 4.2 where it was shown that $v_p v_g = c^2$.

Hence
$$a = 2{\cdot}54 \times 10^{-2}\ \text{m}$$
$$b = 1{\cdot}27 \times 10^{-2}\ \text{m}$$
$$\lambda = c/f = (3 \times 10^8)/10^{10} = 3 \times 10^{-2}\ \text{m}$$
with
$$\lambda_c = 2a = 5{\cdot}08 \times 10^{-2}\ \text{m}$$
Since
$$1/\lambda^2 = 1/\lambda_c^2 + 1/\lambda_g^2$$
We have
$$1/\lambda_g^2 = \frac{1}{(3 \times 10^{-2})^2} - \frac{1}{(5{\cdot}08 \times 10^{-2})^2}$$
$$= 0{\cdot}111 \times 10^4 - 0{\cdot}0388 \times 10^4$$
$$= 0{\cdot}723 \times 10^3$$
or
$$\lambda_g = 3{\cdot}744 \times 10^{-2}\ \text{m}$$
Now
$$v_p = f\lambda_g = 10^{10} \times 3{\cdot}744 \times 10^{-2}$$
or
$$v_p = 3{\cdot}744 \times 10^8\ \text{m/s}$$
with
$$v_g = c^2/v_p = \frac{9 \times 10^{16}}{3{\cdot}744 \times 10^8}$$
or
$$v_g = 2{\cdot}4 \times 10^8\ \text{m/s}$$

4.4 Rectangular waveguides

The properties of waveguides may be studied by solving Maxwell's equations for propagation, in bounded regions. The solutions to Maxwell's equations for rectangular and circular waveguides are obtained by satisfying the necessary boundary conditions for propagation within the waveguide.

The general equation for wave propagation in space is

$$\nabla^2\psi = \frac{1}{c^2}\frac{\partial^2\psi}{\partial t^2}$$

where ψ is the function $\boldsymbol{E}$ or $\boldsymbol{H}$.

Since we are essentially concerned with alternating quantities, which may be represented by $e^{j\omega t}$, hence $\partial^2/\partial t^2 \equiv -\omega^2$

with $$\nabla^2\psi = -(\omega^2/c^2)\psi$$

or $$\nabla^2\psi = -k^2\psi$$

where $k = \omega/c$ gives the eigen-values of the system. Solutions to this equation are given in Jordan[18].

4.5 Rectangular modes

Two types of guided waves are now possible, namely, transverse electric (TE) wave for which $E_z = 0$ and transverse magnetic (TM) wave, for which $H_z = 0$. By using Maxwell's equations for curl $\boldsymbol{E}$ and curl $\boldsymbol{H}$, with the solutions for E_z and H_z, the general field components for the various modes can be obtained. Further details are given in Jordan[18] and the relevant field components are given hereafter where $k_c = 2\pi/\lambda_c$, λ_c is the cut-off wavelength of a given mode and $\gamma \simeq j\beta$.

TE modes

$$E_x = \frac{j\omega\mu H_0}{k_c^2}\frac{n\pi}{b}\cos\frac{m\pi x}{a}\sin\frac{n\pi y}{b}\,e^{j(\omega t - \beta z)}$$

$$E_y = \frac{-j\omega\mu H_0}{k_c^2}\frac{m\pi}{a}\sin\frac{m\pi x}{a}\cos\frac{n\pi y}{b}\,e^{j(\omega t - \beta z)}$$

$$E_z = 0$$

$$H_x = \frac{\gamma H_0}{k_c^2}\frac{m\pi}{a}\sin\frac{m\pi x}{a}\cos\frac{n\pi y}{b}\,e^{j(\omega t - \beta z)}$$

$$H_y = \frac{\gamma H_0}{k_c^2}\frac{n\pi}{b}\cos\frac{m\pi x}{a}\sin\frac{n\pi y}{b}\,e^{j(\omega t - \beta z)}$$

$$H_z = H_0\cos\frac{m\pi x}{a}\cos\frac{n\pi y}{b}\,e^{j(\omega t - \beta z)}$$

The simplest TE mode is $TE_{10}(H_{10})$ or dominant mode which has the longest cut-off wavelength. For the TE_{10} mode this is given by

$$\lambda_c = 2/\sqrt{(1/a)^2} = 2a$$

The field components for the TE_{10} mode are:

$$E_x = 0$$

$$E_y = A \sin\frac{\pi x}{a} e^{j(\omega t - \beta z)}$$

$$E_z = 0$$

with
$$A = \frac{-j\omega\mu H_0}{k_c^2}\pi/a$$

$$H_x = B \sin\frac{\pi x}{a} e^{j(\omega t - \beta z)}$$

$$H_y = 0$$

$$H_z = H_0 \cos\frac{\pi x}{a} e^{j(\omega t - \beta z)}$$

with
$$B = \frac{\gamma H_0}{k_c^2}\pi/a$$

Hence E_y has a half sine wave distribution over the waveguide cross-section with a peak value at the centre of the cross-section where $x = a/2$.

Also
$$E_y/-H_x = A/-B = \frac{\omega\mu}{\beta} = (\omega/c)\mu c(1/\beta) = Z_0(\lambda_g/\lambda)$$

where $Z_0 = \sqrt{\mu_0/\varepsilon_0}$ is the intrinsic impedance of free space. The ratio $E_y/-H_x$ is *defined* as the impedance Z_{TE} for the TE wave.

Hence
$$Z_{TE} = Z_0(\lambda_g/\lambda) = 377\lambda_g/\lambda \text{ ohms}$$

TM modes

$$E_x = \frac{-\gamma E_0}{k_c^2}\frac{m\pi}{a}\cos\frac{m\pi x}{a}\sin\frac{n\pi y}{b} e^{j(\omega t - \beta z)}$$

$$E_y = \frac{-\gamma E_0}{k_c^2}\frac{n\pi}{b}\sin\frac{m\pi x}{a}\cos\frac{n\pi y}{b} e^{j(\omega t - \beta z)}$$

$$E_z = E_0 \sin\frac{m\pi x}{a}\sin\frac{n\pi y}{b} e^{j(\omega t - \beta z)}$$

$$H_x = \frac{j\omega\varepsilon E_0}{k_c^2}\frac{n\pi}{b}\sin\frac{m\pi x}{a}\cos\frac{n\pi y}{b}\,e^{j(\omega t - \beta z)}$$

$$H_y = \frac{-j\omega\varepsilon E_0}{k_c^2}\frac{m\pi}{a}\cos\frac{m\pi x}{a}\sin\frac{n\pi y}{b}\,e^{j(\omega t - \beta z)}$$

$$H_z = 0$$

Similarly, for the TM modes we have

$$E_y/-H_x = \beta/\omega\varepsilon = \frac{\beta c}{\omega\varepsilon c} = \frac{2\pi}{\lambda g}\times\frac{\lambda}{2\pi}\times\frac{1}{\varepsilon c} = Z_0(\lambda/\lambda_g)$$

which is *defined* as the impedance Z_{TM}.

Hence $\qquad Z_{TM} = Z_0(\lambda/\lambda_g) = 377\lambda/\lambda_g$ ohms

The simplest TM mode is the TM_{11} or dominant mode since TM modes vanish when $n = 0$. Typical field patterns for some rectangular modes are shown in Fig. 4.4.

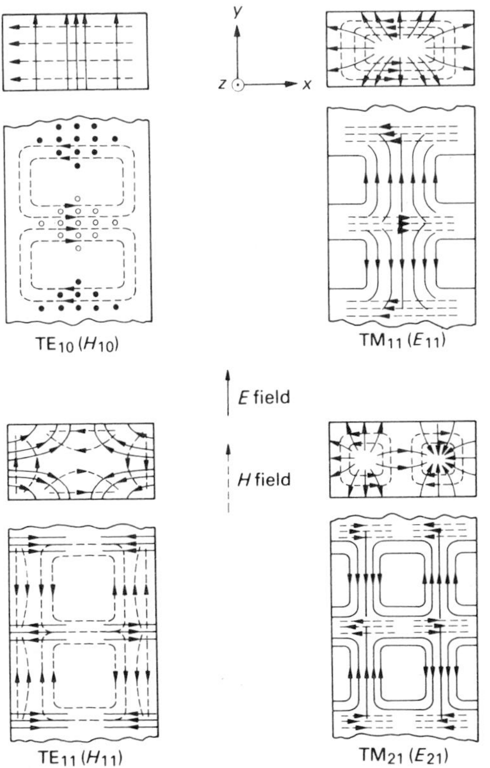

Fig. 4.4

Example 4.2

A rectangular waveguide has internal dimensions a and b as shown in Fig. 4.5. A mode has the electric field given by

$$E_y = A \sin\frac{\pi x}{a} \exp[\mathrm{j}(\omega t - \beta z)]$$

$$E_x = E_z = 0$$

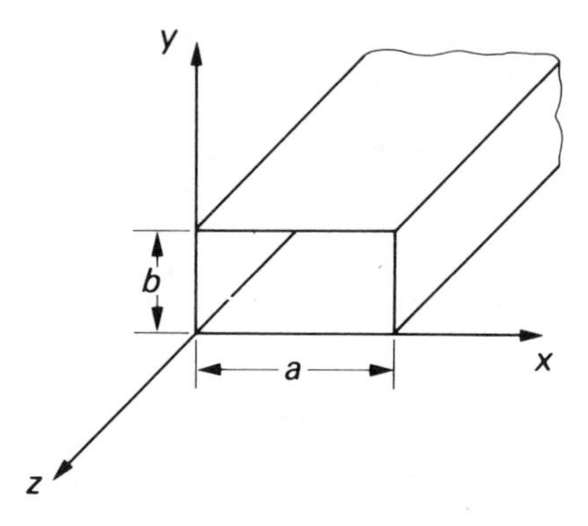

Fig. 4.5

Derive an expression for the phase-change coefficient β in terms of a, ω and the electrical constants of the medium filling the guide. Obtain the corresponding magnetic field components.

Describe how the guide may be coupled to a coaxial line so that the above mode is excited within the guide.

Solution

The mode propagated is the H_{10} where $m = 1$, $n = 0$

Now
$$\gamma = \mathrm{j}\beta \qquad (\alpha \simeq 0)$$

$$\partial/\partial z \equiv -\gamma = -\mathrm{j}\beta$$

$$\partial/\partial t \equiv \mathrm{j}\omega$$

Also
$$\operatorname{curl} \boldsymbol{E} = -\frac{\partial \boldsymbol{B}}{\partial t} = -\mathrm{j}\omega\mu \boldsymbol{H}$$

Hence
$$\frac{\partial E_z}{\partial y} - \frac{\partial E_y}{\partial z} = -\mathrm{j}\omega\mu H_x$$

$$-\frac{\partial E_z}{\partial x} + \frac{\partial E_x}{\partial z} = -\mathrm{j}\omega\mu H_y$$

$$\frac{\partial E_y}{\partial x} - \frac{\partial E_x}{\partial y} = -\mathrm{j}\omega\mu H_z$$

Since $E_z = 0$, we obtain

$$-\frac{\partial E_y}{\partial z} = -j\omega\mu H_x$$

with
$$j\beta A \sin\frac{\pi x}{a} e^{-j\beta z} e^{j\omega t} = -j\omega\mu H_x$$

or
$$H_x = -(\beta A/\omega\mu)\sin\frac{\pi x}{a} e^{-j\beta z} e^{j\omega t}$$

Since $E_x = E_z = 0$ we obtain $H_y = 0$. Also, we have

$$\frac{\partial E_y}{\partial x} = -j\omega\mu H_z$$

with
$$(\pi/a)A\cos\frac{\pi x}{a} e^{-j\beta z} e^{j\omega t} = -j\omega\mu H_z$$

or
$$H_z = (j\pi A/\omega\mu a)\cos\frac{\pi x}{a} e^{-j\beta z} e^{j\omega t}$$

Now
$$1/\lambda_g^2 = 1/\lambda^2 - 1/\lambda_c^2$$

with
$$\lambda_c = \frac{2}{\sqrt{(1/a)^2 + 0}} = 2a$$

$$\omega/c = 2\pi/\lambda$$

$$\beta = 2\pi/\lambda g$$

Hence
$$(\beta/2\pi)^2 = (\omega/2\pi c)^2 - (1/2a)^2$$

or
$$\beta = \sqrt{\omega^2\mu\varepsilon - \pi^2/a^2}$$

The H_{10} mode may be launched by inserting a $\lambda/4$ length of coaxial cable into the side of the guide, placed $\lambda g/4$ from a moveable piston as shown in Fig. 4.6.

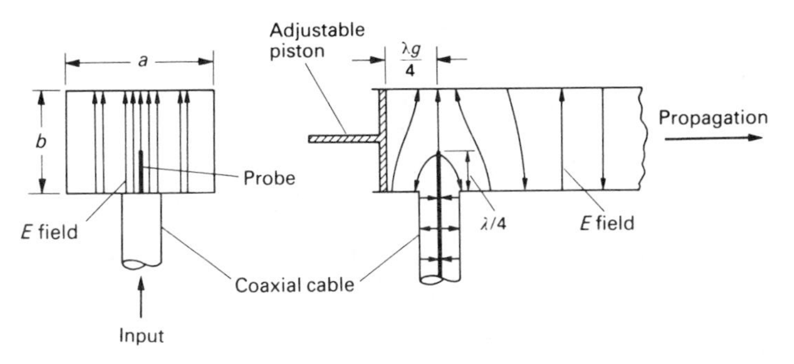

Fig. 4.6

4.6 Circular waveguides

The solution of the wave equation for circular waveguides is conveniently obtained using cylindrical coordinates. Hence we have

$$\nabla^2\psi = -(\omega^2/c^2)\psi$$

where $\psi = \boldsymbol{E}$ or $\boldsymbol{H}$

and

$$\nabla^2 = \partial^2/\partial r^2 + \frac{1}{r}\partial/\partial r + \frac{1}{r^2}\partial^2/\partial\phi^2 + \partial^2/\partial z^2$$

in cylindrical coordinates. Further details are given in Ramo[16].

4.7 Circular modes

The general field equations for these modes are given below and are obtained through Maxwell's curl equations, together with the solutions for $\boldsymbol{E}$ and $\boldsymbol{H}$. Further details are given in Ramo[16].

TE_{mn} modes

$$E_r = \frac{-\omega\mu m H_0}{r k_c^2} J_m(k_c r)\cos m\phi\, e^{j(\omega t - \beta z)}$$

$$E_\phi = \frac{j\omega\mu}{k_c} H_0 J'_m(k_c r)\cos m\phi\, e^{j(\omega t - \beta z)}$$

$$E_z = 0$$

$$H_r = \frac{-\gamma H_0}{k_c} J'_m(k_c r)\cos m\phi\, e^{j(\omega t - \beta z)}$$

$$H_\phi = \frac{-\beta m H_0}{r k_c^2} J_m(k_c r)\cos m\phi\, e^{j(\omega t - \beta z)}$$

$$H_z = H_0 J_m(k_c r)\cos m\phi\, e^{j(\omega t - \beta z)}$$

TM_{mn} modes

$$E_r = \frac{-\gamma E_0}{k_c} J'_m(k_c r)\cos m\phi\, e^{j(\omega t - \beta z)}$$

$$E_\phi = \frac{-\beta m E_0}{r k_c^2} J_m(k_c r)\cos m\phi\, e^{j(\omega t - \beta z)}$$

$$E_z = E_0 J_m(k_c r)\cos m\phi\, e^{j(\omega t - \beta z)}$$

$$H_r = \frac{\omega\varepsilon m E_0}{r k_c^2} J_m(k_c r)\cos m\phi\, e^{j(\omega t - \beta z)}$$

$$H_\phi = \frac{-j\omega\varepsilon E_0}{k_c} J'_m(k_c r)\cos m\phi\, e^{j(\omega t - \beta z)}$$

$$H_z = 0$$

Comments

1. $E_r/H_\phi = \omega\mu/\beta = 377\,\lambda g/\lambda = Z_{TE}$
2. $E_r/H_\phi = \beta/\omega\varepsilon = 377\,\lambda/\lambda g = Z_{TM}$

Typical field patterns for some TE and TM modes are shown in Fig. 4.7.

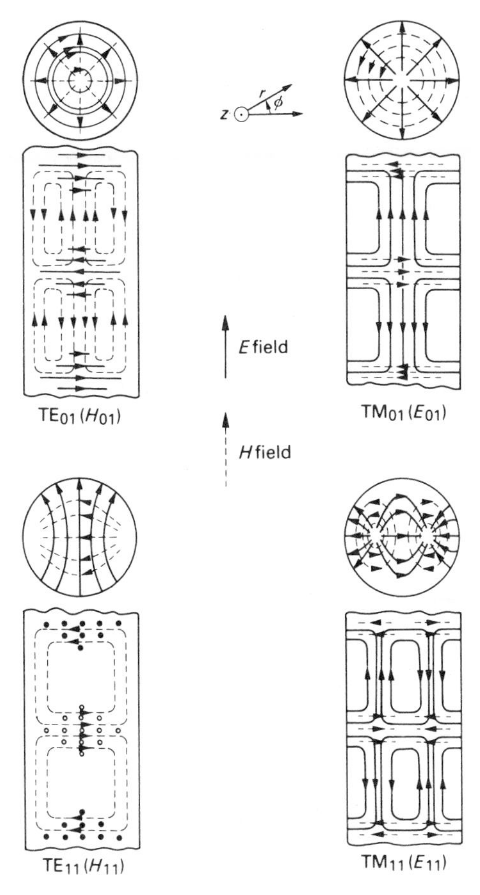

Fig. 4.7

4.8 Higher-order modes[22]

The principal or dominant mode in a rectangular waveguide is the TE_{10} mode ($m = 1$ and $n = 0$), which has the longest wavelength of propagation. It is the most commonly used waveguide mode, because of its minimum attenuation and ease in launching. Modes which have shorter wavelengths are called *higher-order* modes and are designated with various other values of the subscripts m and n such as TE_{20}, TM_{11} etc. These higher-order modes are not normally used (except in certain applications) because of their higher attenuation and more complex launching requirements.

Like the dominant mode, each higher-order mode is associated with its own cut-off wavelength λ_c, and guide wavelength λ_g, which are given by the general expressions

$$\lambda_c = \frac{2}{\sqrt{(m/a)^2 + (n/b)^2}}$$

and

$$\lambda_g = \frac{\lambda}{\sqrt{1 - (\lambda/\lambda_c)^2}}$$

where m, n are the subscripts mentioned earlier, a, b are the guide dimensions and λ is the wavelength of propagation. The propagation of any particular higher-order mode is therefore largely dependent on the choice of guide dimensions and on the wavelength of propagation. In practice, the higher-order modes are avoided by choosing the right guide dimensions and operating frequency for the dominant TE_{10} mode.

Higher-order modes can also be propagated in circular waveguides. The principal mode which has the longest wavelength (analogous to the rectangular TE_{10} mode) is the TE_{11} mode which is used most commonly. In general, circular waveguide modes have higher attenuation than the rectangular modes, with the exception of the TE_{01} mode which has the lowest attenuation, and the attenuation decreases with an increase in frequency. It is therefore particularly attractive for use at very high frequencies. In the past, research work has been undertaken to exploit its use for various applications using millimetric waves around 90 GHz.

Each higher-order circular mode is also associated with its own cut-off wavelength λ_c and guide wavelength λ_g. The former, however, is related by a complex expression in terms of the Bessel functions J_0, J_1 etc. and the radius r of the circular waveguide. The guide wavelength λ_g is given by the expression

$$\lambda_g = \frac{\lambda}{\sqrt{1 - (\lambda/\lambda_c)^2}}$$

where λ is the wavelength of propagation and λ_c is the cut-off wavelength of the mode.

A particular circular mode of interest in radar applications is the TM_{01} mode which is used for rotating joint structures and the TE_{11} mode which is used in the construction of the rotary vane attenuator. Table 1.2 in Chapter 1 lists some cut-off wavelengths for certain rectangular and circular modes.

Example 4.3

Draw the field patterns for the H_{01} mode in a circular waveguide and explain why this mode was chosen for trunk telecommunications applications. Estimate the critical wavelength for this mode and show that it is not the dominant mode in circular waveguide. How can the waveguide be constructed to inhibit the propagation of the other modes? (C.E.I.)

Solution

The H_{01} mode is illustrated in Fig. 4.8 and it was chosen for long-distance trunk communications because it has very low losses which tend to zero at high frequencies. Typically, the attenuation at 16 GHz is about 0·005 dB/m in a 5 cm diameter copper wall circular waveguide, which is the lowest attenuation for any circular mode.

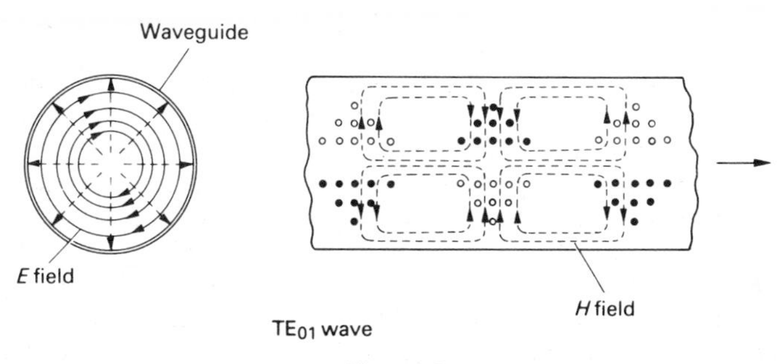

Fig. 4.8

The critical wavelength for this mode is $1{\cdot}64\,a$, where a is the radius of the waveguide. By using a larger radius (oversize waveguide) than that given by the cut-off condition, attenuation can be made small and it becomes attractive for long-distance communication. On the other hand, the dominant circular mode is the TE_{11} (H_{11}) mode with the longest cut-off wavelength of $3{\cdot}42\,a$, but its losses are much greater.

However, a disadvantage of oversize waveguide is the propagation of spurious modes with consequent loss of power. Recent work undertaken has led to the suppression of unwanted modes by making them travel more slowly along the waveguide than the H_{01} mode, thus reducing energy transfer to these modes. This is done by using either a circular waveguide made of a wire helix or a metal surface which has corrugations or a thin coating of dielectric material. Mode filters may also be employed to clean up any remaining modes.

Comment

Trunk waveguide systems have now been superseded by fibre-optic systems.

4.9 Waveguide attenuation

When an electromagnetic wave travels down a hollow waveguide, currents are induced in the walls of the waveguide by the tangential magnetic field component. This gives rise to ohmic losses on the inside walls and so the wave is gradually attenuated as it travels along the waveguide. The attenuation is a function of the waveguide material, dimensions, wavelength and the mode of propagation which determines the strength of the tangential component of magnetic field H. To reduce losses, the inner surface of the waveguide may be plated with a thin layer of silver or gold.

Calculations of losses can be made from theoretical reasoning using the field equations and it leads to an expression for the average power loss P_L which is given by

$$P_L = \tfrac{1}{2}|H_t|^2 R_s \text{ watts/m}^2$$

where H_t is the tangential magnetic field component near the internal wall surface and R_s is the resistivity of the surface. Losses are then calculated for each inner surface and summed to give the total losses per unit length of waveguide. From these losses, the attenuation coefficient α per unit length of waveguide is evaluated and it can be expressed in decibels per metre. Typical curves of waveguide losses are given in Fig. 4.9 from which it will be seen that losses are greater for circular modes as compared to rectangular modes. In practice, circular mode losses are greater than the theoretical values because of the difficulty of making a perfectly circular waveguide.

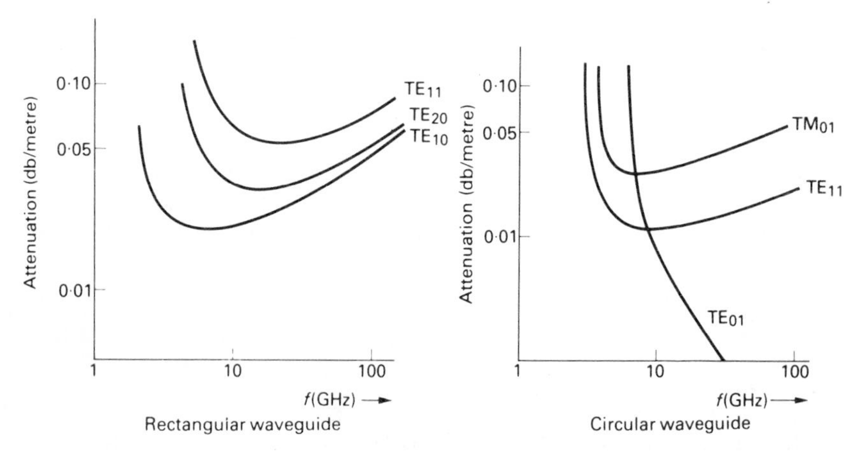

Fig. 4.9

The mode with the minimum attenuation is generally used and for a rectangular waveguide this is the TE_{10} or dominant mode which typically has an attenuation of about 0·06 dB/m at 3 GHz or 0·03 dB/m at 10 GHz. In circular waveguide, the principal mode of practical interest is the TE_{11} mode which typically has an attenuation of 0·2 dB/m at 3 GHz and 0·01 dB/m at 10 GHz.

Of particular interest is the TE_{01} mode in circular waveguide which has an exceptionally low attenuation and which tends to zero at very high frequencies. A practical application of this is the possibility of transmitting microwave energy at millimetric wavelengths. Typically, a circular waveguide of 2 inch diameter has an attenuation of less than 3 dB/km over the frequency range 30–110 GHz.

The attenuation losses described so far are for the *propagating* modes which operate below the cut-off wavelength λ_c for a given waveguide. However, if a waveguide is operated *above* the cut-off wavelength λ_c, the wave is greatly attenuated and does not propagate down the waveguide. These 'evanescent' modes are found to exist in the form of induction fields which travel back and forth *across* the waveguide with reactive losses. They are attenuated exponentially with distance and the attenuation can be calculated from the field equations. When expressed in decibels, the attenuation α is given by the expression

$$\alpha = \frac{54{\cdot}6}{\lambda_c} \sqrt{1 - (\lambda_c/\lambda)^2}$$

where λ_c is the cut-off wavelength and λ is the operating wavelength. If $\lambda \gg \lambda_c$, this reduces to the value $\alpha \simeq 54{\cdot}6/\lambda_c$ which is virtually independent of the operating wavelength λ.

One application of these evanescent modes is in the design of a 'piston' attenuator. The attenuator consists of two circular sections of waveguide which slide one within the other. If energy is coupled into one section by a loop, it can be coupled out of the section by another loop. The output energy is then a function of the distance between the two loops and it can be directly calibrated in decibels.

Example 4.4

Derive expressions for the total power flow P_T and characteristic impedance Z_0 of a rectangular waveguide propagating a TE_{10} mode in terms of its dimensions a and b. If such a waveguide has $a = 3$ cm, $b = 1{\cdot}5$ cm, determine the values of P_T and Z_0 when the maximum voltage across the waveguide is 1 kV at an operating frequency of 10 GHz.

Solution

The power density in the waveguide is given by

$$P = E \times H \text{ watts/m}^2$$

and for a TE_{10} wave travelling in the *z-direction* as in Fig. 4.5 we obtain

$$P = E_y \times (-H_x) \text{ watts/m}^2$$

where

$$E_y = K \sin\left(\frac{\pi x}{a}\right) e^{j(\omega t - \beta z)}$$

$$H_x = -\frac{K\beta}{\omega\mu} \sin\left(\frac{\pi x}{a}\right) e^{j(\omega t - \beta z)}$$

$$K = \frac{V_{max}}{b}$$

and the *real* power density in the z-direction is given by

$$P = \frac{K^2\beta}{\omega\mu}\sin^2\left(\frac{\pi x}{a}\right)\cos^2(\omega t - \beta z)$$

with
$$P_{\text{av.}} = \frac{K^2\beta}{2\omega\mu}\sin^2\left(\frac{\pi x}{a}\right)\text{ watts/m}^2$$

The total power flow P_T down the waveguide is obtained by integrating over the cross-section of the waveguide and we obtain

$$P_T = \int_0^a\int_0^b P_{\text{av.}}\,\mathrm{d}x\,\mathrm{d}y = \frac{K^2\beta}{2\omega\mu}\int_0^a\int_0^b \sin^2\left(\frac{\pi x}{a}\right)\mathrm{d}x\,\mathrm{d}y$$

or
$$P_T = \frac{K^2\beta ab}{4\omega\mu}\text{ watts}$$

Also, if the rms voltage across the waveguide is V_{rms} and its characteristic impedance is Z_0, we have

$$P_T = V_{\text{rms}}^2/Z_0$$

or
$$Z_0 = V_{\text{rms}}^2/P_T = V_{\text{max}}^2/2P_T$$

where
$$V_{\text{max}} = |E_y|b = Kb$$

Hence
$$Z_0 = \frac{4K^2b^2\omega\mu}{2K^2\beta ab}$$

or
$$Z_0 = \frac{2\omega\mu b}{a\beta}\text{ ohms}$$

Problem

For $V_{\text{max}} = 1{\cdot}5$ kV and $b = 1{\cdot}5$ cm we obtain

$$K = 1{\cdot}5 \times 10^3/1{\cdot}5 \times 10^{-2} = 10^5\text{ V/m}$$

$$\beta = \sqrt{\omega^2/c^2 - (\pi/a)^2}$$

$$= \sqrt{(4\pi^2 \times 10^{20}/9 \times 10^{16}) - (\pi/3 \times 10^{-2})^2}$$

$$= \sqrt{4{\cdot}3876 \times 10^4 - 1{\cdot}097 \times 10^4}$$

$$= \sqrt{3{\cdot}2906 \times 10^4}$$

or
$$\beta = 181{\cdot}4\text{ rad/m}$$

Thus
$$P_T = \frac{10^{10} \times 181{\cdot}4 \times 3 \times 10^{-2} \times 1{\cdot}5 \times 10^{-2}}{4 \times 2\pi \times 10 \times 10^9 \times 4\pi \times 10^{-7}}$$

$$= \frac{181{\cdot}4 \times 4{\cdot}5 \times 10^3}{8 \times 4 \times \pi^2}$$

or $$P_T = 2{\cdot}58\ \text{kW}$$

Also $$Z_0 = \frac{2 \times 2\pi \times 10 \times 10^9 \times 4\pi \times 10^{-7} \times 1{\cdot}5 \times 10^{-2}}{3 \times 10^{-2} \times 181{\cdot}4}$$

$$= \frac{8\pi^2 \times 10^3}{181{\cdot}4}$$

or $$Z_0 = 435\ \text{ohms}$$

4.10 Launching in waveguides

Waves may be launched into a waveguide from a coaxial cable using electrostatic coupling, by means of a short probe antenna, or by using electromagnetic coupling, by employing a small closed loop antenna.

If the inner conductor of a coaxial cable is inserted as a $\lambda/4$ probe into the waveguide and $\lambda g/4$ from one end of the waveguide, waves are propagated towards the right as shown in Fig. 4.10(a) and the end position of the waveguide is made adjustable by means of a moveable plunger. The field pattern near the probe contains higher-order modes but further down the waveguide, a single mode is transmitted.

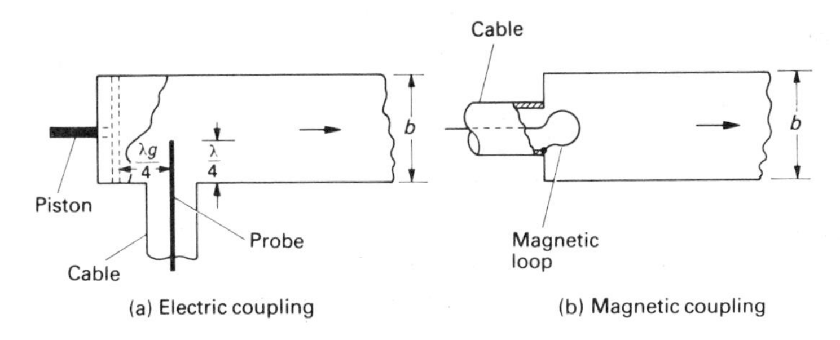

Fig. 4.10

In the case of magnetic coupling, which is shown in Fig. 4.10(b), the inner conductor of a coaxial cable is bent around in the form of a closed loop. The size of the loop is not critical and it couples into the waveguide a set of magnetic field lines. The end of the loop is terminated on the waveguide and it behaves as a single-turn coil antenna.

Various types of modes may be launched into a waveguide depending upon the type of coupling used and on its position inside the waveguide.

Typical launching techniques for a rectangular waveguide are illustrated in Fig. 4.11 for three waveguide modes.

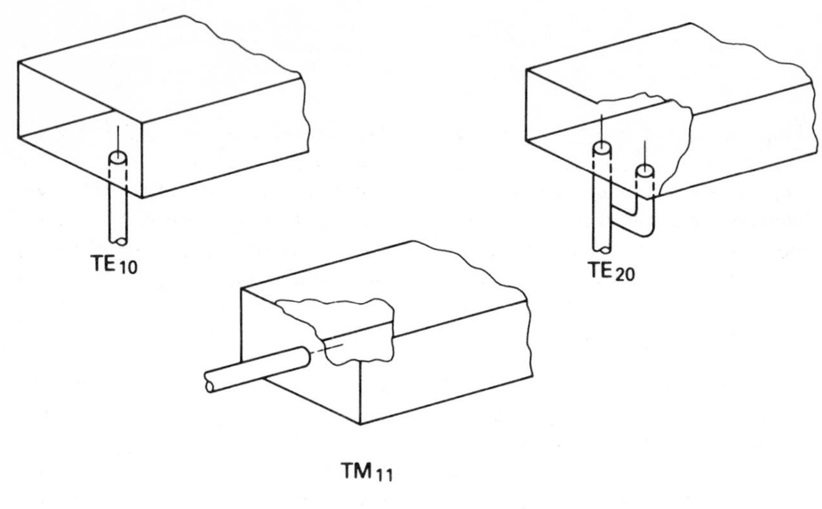

Fig. 4.11

5
Microwave techniques

The increasing use of microwaves, because of demands for higher frequencies due to the congestion at lower frequency bands, has led to considerable development in this field. Microwave engineering is now a firmly established branch of telecommunications and a broad knowledge of the various aspects pertinent to this field is essential. These aspects may be broadly classified under sources, components and measurements.

5.1 Microwave sources[23, 24]

The main source of low-power microwave energy in the past has been the klystron oscillator. Two basic forms are the two-cavity klystron and the reflex klystron, the latter being the most commonly employed in previous years. More recently, solid state sources like the Gunn diode oscillator are now widely used in low-power applications. For high-power applications, the multicavity magnetron, travelling wave amplifier (TWA) and crossed-field amplifier (CFA) have found considerable use in radar systems.

(a) Two-cavity klystron

In the two-cavity klystron shown in Fig. 5.1(a), an electron source called the *cathode* emits a beam of electrons which are accelerated along a glass tube by a positive potential applied to the anode or *collector* at the opposite end. Electrons are first accelerated across a gap which is coupled to a resonant cavity called the *buncher*. They are subsequently allowed to drift through a *drift space* before entering a second gap which is coupled to another resonant cavity called the *catcher* and are finally collected by the collector. The electron stream is focussed axially by a focussing coil or by permanent magnets placed around the glass tube. The device can be operated as an amplifier or as an oscillator, but it is normally used as a power amplifier at high frequencies because of the short transit time of the accelerated electrons.

The operation of the klystron depends on the principle of velocity modulation, whereby the uniform beam of electrons leaving the cathode is subsequently made to form 'bunches' of electrons, an effect known as *bunching*. To understand how this bunching action arises, consider the Applegate diagram shown in Fig. 5.1(b) which is a plot of distance x along the tube against the time of arrival of the moving electrons. Thus, the slope

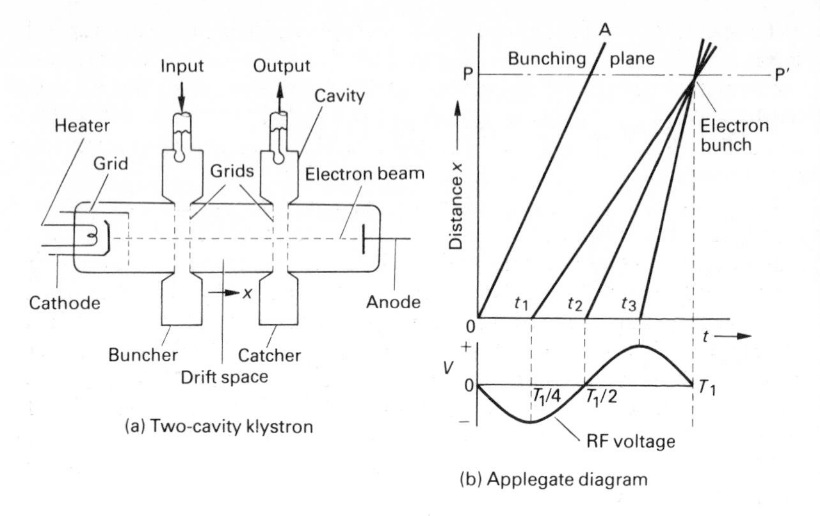

Fig. 5.1

of any line such as 0A represents the *velocity* of a moving electron along the tube. In Fig. 5.1(b), electrons are shown arriving at the first gap at the time instants 0, t_1, t_2, t_3 of the RF voltage which is either applied to this gap (amplifier case) or is assumed to exist initially at the first gap due to random noise fluctuations present in the stream of moving charge (oscillator case).

At time t_1, the electron entering the first gap is decelerated by the negative RF voltage and its velocity is decreased slightly, while at time t_2, an electron moves through the first gap when the RF voltage is zero, and it passes through with its initial uniform velocity. At time t_3, however, the electron entering the first gap is accelerated by the positive RF voltage and its velocity is slightly increased. The individual electron velocities are given by the *slopes* of the lines drawn at time instants t_1, t_2, t_3 respectively. Thus, it is observed that these lines meet at a point and the corresponding electrons tend to arrive in a bunch at some distance beyond the first gap. This grouping of electrons at the bunching plane PP' shows that the electrons are made to bunch at various times such as T_1 for the first group, T_2 for the second group etc.

The bunches of electronic charge are allowed to drift through a field-free space known as the 'drift space' and during this period faster moving electrons catch up with slower moving electrons emitted earlier in the RF cycle. Hence, the electrons are capable of exchanging energy amongst themselves and move forward together as a bigger bunch. Bunching action is therefore increased in the drift space and subsequently the bigger charge bunches pass through the second gap. As this gap is coupled to another resonant cavity, the moving bunches induce a large voltage across it and are finally collected at the final anode or collector.

To operate the device as an amplifier, a magnetic loop is used to couple the input signal into the first cavity and a similar loop is used for extracting the amplified signal from the second cavity. Alternatively, if some energy is fed back from the catcher to the buncher to sustain the initial RF voltage across the first gap, the device functions as an oscillator and continuous wave (CW) power can be extracted from the second cavity.

For practical applications, the output power of the two-cavity klystron is insufficient. It can be increased by using a multicavity klystron consisting of 3 or 4 cavities in tandem. This multicavity structure improves the bunching action and increases output power considerably. Typical multicavity klystron amplifiers are used for television broadcast or radar systems. They are capable of producing CW powers around 50 to 100 kW at frequencies between 400 and 800 MHz or pulsed powers of 10 to 40 MW peak at frequencies between 1 and 10 GHz. Frequency bandwidths are generally narrow and around 1–2%, but recently, wider bandwidths approaching 10% have been attained for power applications in satellite ground stations or troposcatter systems.

(b) Gunn diode oscillator[25]

In recent years, a more convenient form of microwave oscillator which uses a low-voltage d.c. supply instead of the cumbersome high-voltage supply of the klystron, is the Gunn diode oscillator shown in Fig. 5.2. In certain semiconductors such as *n*-type GaAs or GaP, oscillations can occur in bulk material rather than across a junction. This is known as the *Gunn effect*. In a Gunn diode device, the current increases linearly with applied voltage up to a certain threshold value, after which it decreases with increasing voltage thus producing a negative resistance effect.

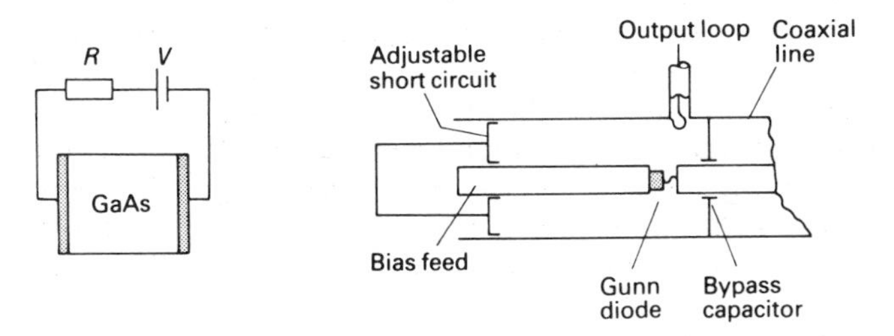

Fig. 5.2

A typical Gunn diode consists of a substrate of GaAs to which an electrode is connected (the cathode). A thin epitaxial layer of GaAs is formed on its surface to which another electrode is connected (the anode). To produce an oscillator, the diode may be mounted in a coaxial cavity as shown in Fig. 5.2. The cavity is tuned at one end by a moveable piston and energy is coupled out by a magnetic loop at the other end. Since the GaAs sample is very thin, a few volts of d.c. bias is sufficient to produce current pulses through the sample which shock excite the cavity and produce oscillations if the cavity is correctly tuned.

The Gunn effect is due to conduction band electrons which can exist in two energy states. In the low-energy state, they have high mobility and low effective mass, while in the high-energy state, they have low mobility and a high effective mass. The effective mass m_e^* of the electron is the apparent mass with which the electron appears to move when it is under the influence of the wave field of the crystal structure. It can be smaller or greater than the physical mass m_e of the electron and is given by

$$m_e^* = \hbar^2/(d^2W/dk^2)$$

where $\hbar = h/2\pi$ and h is Planck's constant, W is the energy of the electron and k is the wave number.

Physically, the negative resistance effect can be explained by the presence of two conduction band valleys which are separated by a small energy gap of about 0·36 eV as shown in Fig. 5.3. Energy which is gained from the applied electric field causes electrons in the lower conduction band valley to be transferred to the higher conduction band valley, around applied fields of several kV/cm. However, in the higher energy state, the effective mass of the electron is higher and its mobility is lower. Thus, lower mobility implies lower current flow with increasing voltage and it gives rise to the negative resistance effect.

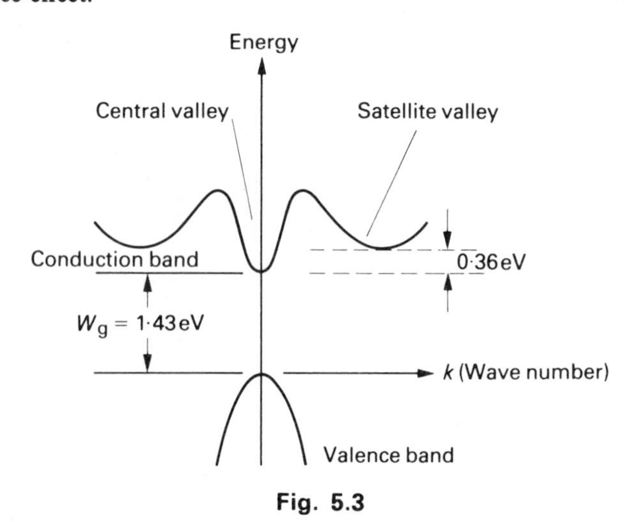

Fig. 5.3

Furthermore, it is also observed that part of the material has a high field gradient across it while the other part has a low field gradient across it. The high field region or *domain* consisting of a cluster of electrons travels from the cathode with a drift velocity of about 10^7 ms^{-1} to the end of the material and collapses at the anode, while another domain is generated at the cathode. The sudden collapse of the domain generates a current pulse and if the transit time through the material is such that the current pulses occur at a microwave frequency, microwave power generation is possible in an external circuit. For this to occur, the GaAs sample must be very thin (about 10 μm) and it must be placed inside a resonant cavity which may be tuned as mentioned earlier.

The description given above is for the Gunn effect *mode* and the frequency of oscillations is almost independent of the external circuit because it is governed mainly by the transit time through the GaAs sample. However, by using the LSA mode (limited space-charge accumulation), the frequency of the oscillation can be controlled to a large extent by the external circuit. Typical frequency ranges cover the *X*-band (8–12 GHz) with bandwidths of about 10%. Both CW and pulsed devices are available with power levels from several milliwatts (CW) to about 100 W (pulsed) while power efficiencies are between 2% and 4% (CW) or 10% and 25% (pulsed). More recently, the frequency of operation of these devices has been extended to about 100 GHz.

5.2 Microwave components[26, 27]

The quality and precision of microwave components is of prime importance in making accurate measurements and it depends on good engineering practice and the tolerances employed. Enumerated below are some typical waveguide components with a short description of each. Details of some ferrite components are given in Appendix B.

Bends and twists

Most waveguides are manufactured in straight sections. However, to go around corners or bends, *E*-plane or *H*-plane bends are used over short lengths. Furthermore, by means of a 90° twist, the plane of polarisation can be changed from horizontal to vertical as illustrated in Fig. 5.4.

T-junctions

To provide additional branches, T-junctions are used. These are either of the *E*-plane or *H*-plane construction or a combination of both which is known as the hybrid or magic-T and is shown in Fig. 5.5. In the case of the *H*-plane T-junction and the *E*-plane T-junction of Fig. 5.5, power entering arm A divides equally into arms B and C if they are properly matched. The configurations are essentially three-port networks. The hybrid junction shown in Fig. 5.5 is a four-port network and is such that power into arm A divides equally into arms B and C if the network is matched. Similarly, power into arm D divides equally but in *antiphase* into arms B and C if the network is matched. Hence, arms A and D are completely isolated from one

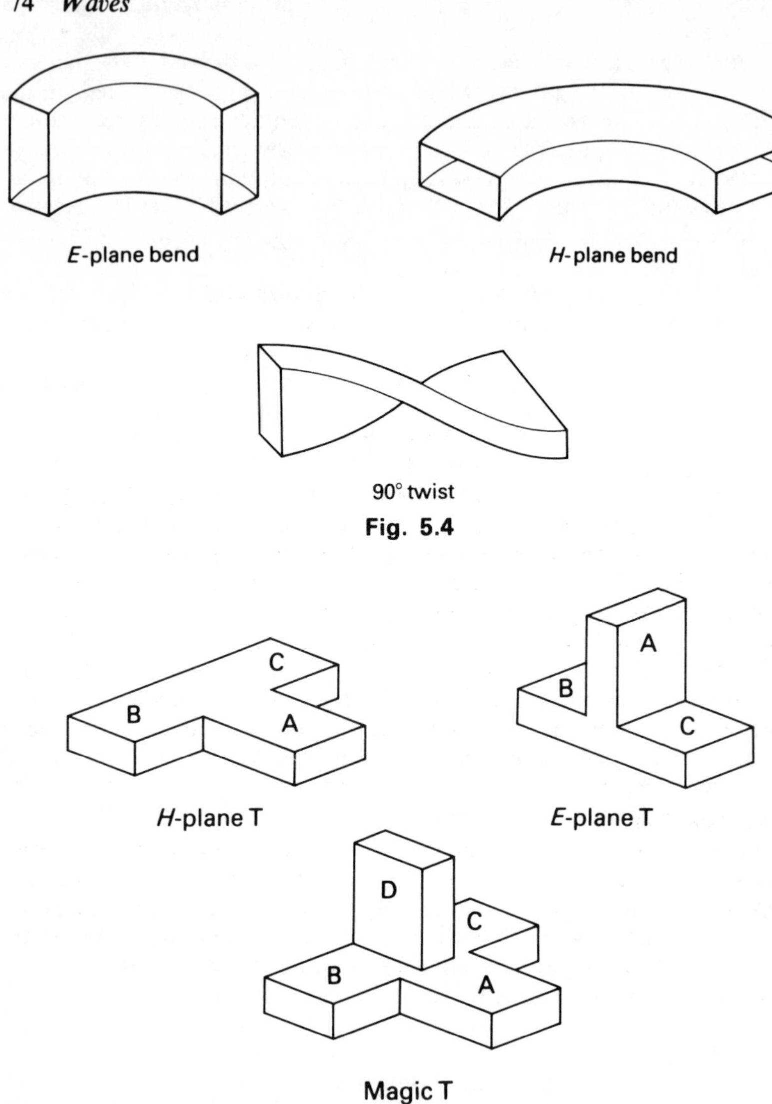

Fig. 5.4

Fig. 5.5

another. The component is often used in a balanced mixer for radar systems.

Hybrid-ring

A variation of the four-port T-junction is the hybrid-ring shown in Fig. 5.6. It consists of an *E*-plane ring around which are four arms A, B, C and D. The

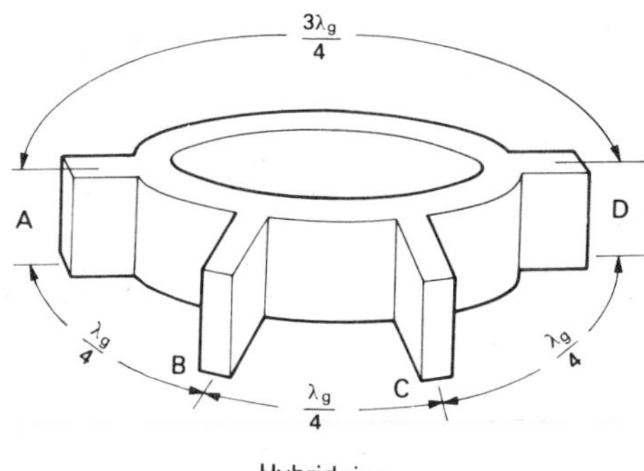

Hybrid-ring

Fig. 5.6

arms are spaced 60° or $\lambda_g/4$ apart and around the ring. As arms A and D are 180° or $3\lambda_g/4$ apart, they are physically opposite to one another.

If power enters arm A, equal powers (but 180° out of phase) are obtained from arms B and D, but nothing from arm C. Similarly, if power enters arm C, equal powers (and in phase) emerge from arms B and D, but nothing from arm A. Thus, arms A and C are isolated from one another.

Usually, the thickness of the central ring and the widths of the side arms are chosen to give a broadband impedance match. Typically, if Z_{0r} is the ring impedance and Z_{0p} is a port impedance, good performance is obtained if $Z_{0p} = \sqrt{2}\, Z_{0r}$ over a 7% bandwidth.

Directional coupler

Typically, this component is a three-port matched network in which power is fed into one port and emerges out of the other two ports. Two important properties of this component are its *coupling* factor and *directivity* factor. For the three-arm coupler shown in Fig. 5.7, power in at arm A appears at arms B and C, while power in at arm B appears at arms A and C.

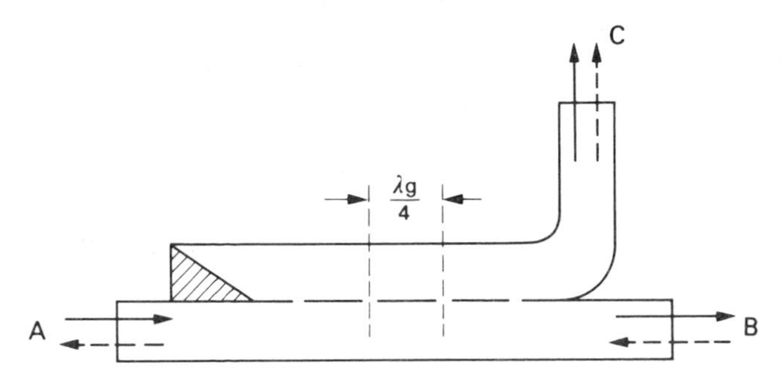

Fig. 5.7

The coupling of power into arm C from arm A is achieved by using carefully machined slots in precision type components spaced $\lambda_g/4$ apart. The basic method of operation depends on the distances between the slots which are such that waves in the forward direction are in phase, while waves in the opposite direction are out of phase and cancel out. One end is usually closed with a matched termination to absorb any reflected power.

For the three-arm coupler shown in Fig. 5.7 with power P_A into arm A and power P_C from arm C (with arm B matched), the coupling factor is defined as

$$\text{Coupling} = 10 \log_{10} (P_A/P_C) \text{ dB}$$

With power P_B into arm B and power P_C from arm C (with arm A matched), the directivity factor is defined as

$$\text{Directivity} = 10 \log_{10} (P_B/P_C) \text{ dB}$$

Coupling factors may vary from 3 dB to 60 dB while a directivity factor is usually better than 60 dB in a precision component.

Attenuators

For general purpose use, a vane attenuator may be employed to attenuate power flowing down a waveguide. The attenuator consists of a vertical glass vane coated with a resistive material like nickel or chromium. It is inserted in the path of the wave, parallel to the narrow wall of the waveguide, to a variable depth by means of a screw mechanism. The greater the depth, the greater is the attenuation and it can be calibrated in decibels.

For higher precision, a rotary vane attenuator is used. It consists of a central circular piece of waveguide with rectangular ends, each end being provided with a resistive vane, while the central part is rotatable as shown in Fig. 5.8.

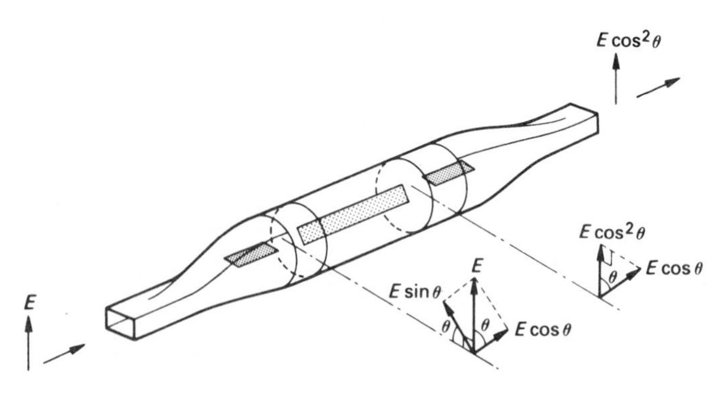

Fig. 5.8

Resolving the E-vector along and perpendicular to the vane, the component $E \sin\theta$ is attenuated, while the component $E \cos\theta$ is transmitted. At the next vane, this vertical component becomes $E \cos^2\theta$ and is transmitted to the other end. Thus, the power finally emerging is proportional to $\cos^2\theta$ and the attenuation is given in decibels by

$$\text{Attenuation} = 20 \log_{10}\left[\frac{E}{E\cos^2\theta}\right] = 40 \log_{10} \sec\theta$$

The angular rotation of the central portion can be calibrated directly in dB with little backlash and with an accuracy of about 0·005 dB.

Wavemeters

These instruments are of two types namely, the transmission wavemeter and the reaction or absorption wavemeter. The reaction wavemeter is generally used because it is placed across the waveguide and can be isolated from it without dismantling the waveguide, by simply detuning the wavemeter. When tuned precisely to the signal frequency, it absorbs power and the power absorption is easily observed on a power meter.

Operation is usually in the TE_{011} cavity-mode and consists of (a) a waveguide resonant cavity, (b) a tuning plunger and (c) a screw mechanism scaled in divisions. The divisions correspond to various frequencies over a limited microwave band e.g. X-band, and they can be read from a calibration chart provided by the manufacturer.

The resonant cavity is very precisely machined, is cylindrical in shape and is gold-flashed for a high-Q. It is mounted on a piece of rectangular waveguide and the cavity size is altered by moving the tuning plunger which is attached to the screw mechanism. Further details of cavity resonators are given in Appendix C.

Standing wave detector

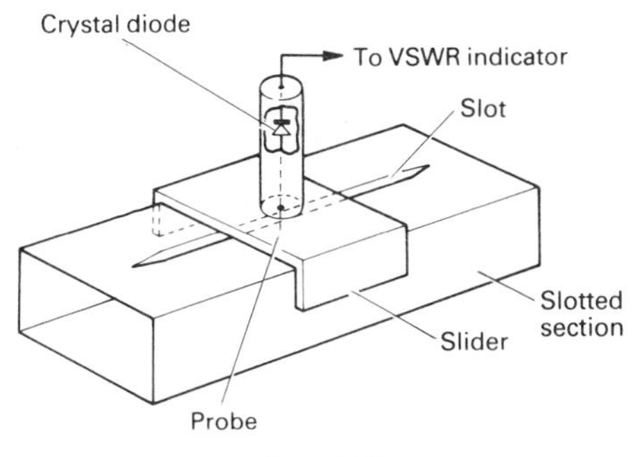

Fig. 5.9

This device is probably the most important and expensive component in a precision waveguide test-bench. Essentially, it is a section of waveguide with a narrow horizontal slot on top as shown in Fig. 5.9. Mounted above is a moving mechanism with a vertical, adjustable probe which is inserted partially into the slot. The whole unit is extremely accurately machined and carefully mounted for precision. It is used to measure the *E*-field along the slot, over a few wavelengths and, from this measurement, the VSWR, reflection coefficient and general mismatch of the waveguide can be determined.

Example 5.1

A hybrid T-junction consists of a length of rectangular waveguide AB with shunt limb C and a series limb D. Draw a sketch of this device and label each limb.

Assuming that A and B are terminated in reflectionless loads, explain by means of separate diagrams showing the *E*-field patterns, why no energy emerges

(a) from D, when the hybrid is energised via limb C
(b) from C, when the hybrid is energised via limb D.

Solution

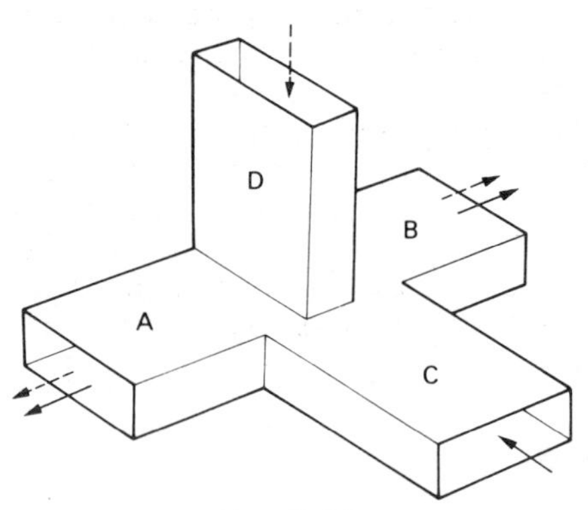

Fig. 5.10

The sketch is shown in Fig. 5.10.

(a) Power fed into limb C divides equally into limbs A and B due to symmetry, and is in phase. No power therefore emerges from limb D which is thus isolated from limb C. This is shown in Fig. 5.11(a).
(b) Power fed into limb D divides equally into limbs A and B due to symmetry, but in opposite phase. No power therefore emerges from limb C which is thus isolated from limb D. This is shown in Fig. 5.11(b).

5.3 Microwave measurements[28, 29]

In the manufacture, installation and operation of microwave equipment, it is necessary to make precise measurements at various microwave frequencies. Such measurements are somewhat more difficult to perform than those at

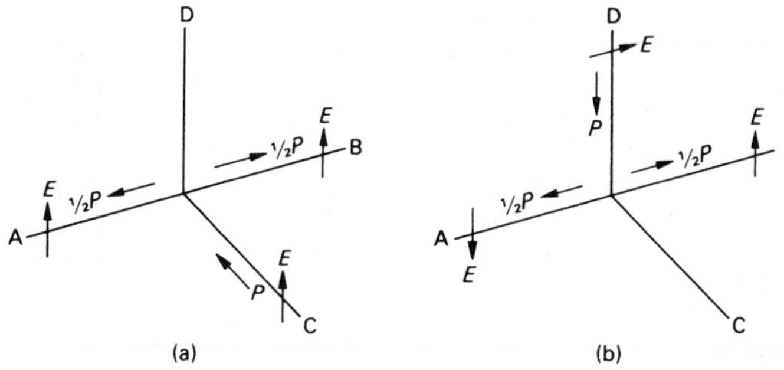

Fig. 5.11

radio frequencies because microwave frequencies are high up in the gigahertz range, which for *X*-band, lies between 8·2 GHz and 12·4 GHz. Such high frequency measurements call for special techniques and precautions.

The measurements usually involve a typical test-bench set-up as shown in Fig. 5.12, comprising for example, grade 2 equipment for general purpose measurements or grade 1 equipment for precision measurements. The selection of proper equipment and measuring techniques depends primarily on the accuracy, speed and costs required by the application. Some applications require a knowledge of both amplitude and phase characteristics of microwave components and are usually made in design laboratories. The bulk of microwave measurements made in production, maintenance and calibration require only the amplitude characteristics of the components tested.

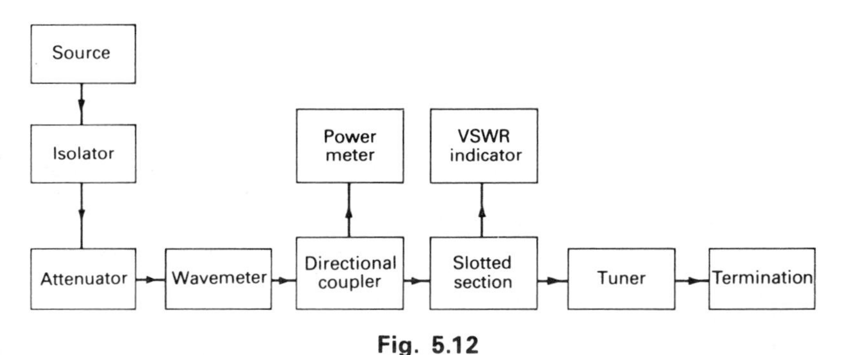

Fig. 5.12

In the past, it was usual to make measurements at spot frequencies. This generally used narrowband equipment and was very time consuming. At present, swept frequency techniques are used, which call for broadband characteristics and have obvious advantages in speed and convenience when testing-time and costs are important considerations. Microwave measurements may also be made with the aid of a computer which is able to store data, program the measurements and make any corrections which may be required in the results.

In Fig. 5.12, the source of microwave power is usually a solid state source if measurements are made at *X*-band. It is followed by an isolator which prevents any reflections from affecting the source stability. The attenuator (general purpose or rotary vane type), sets the power level in the waveguide to a suitable value and the wavemeter is tuned to check the frequency and is then detuned. A directional coupler may be used to divide the power into two branches and a known proportion of the incident power such as half of it (3 dB coupler), is coupled to one arm and monitored by a suitable power meter.

The rest of the power passes to the slotted line which keeps a check on the VSWR. The output of the slotted line probe is fed into the VSWR indicator which reads VSWR directly. The power down the main waveguide is then fed via a three stub tuner and ends in the termination which may be a short circuit, an open circuit, a matched load or any microwave component under test. The three-stub tuner may be used when better matching conditions are required. Furthermore, for reflectometer measurements, the slotted line can be replaced by two three-arm directional couplers placed back-to-back, one for measuring the incident power and the other for measuring the reflected power.

Frequency measurements

The measurements of a microwave frequency often require great accuracy. Frequency measuring devices are either mechanical devices, in which physical dimensions are used to determine frequencies such as cavity wavemeters and slotted lines, or they are electronic devices, in which frequency is determined electronically using heterodyne methods and frequency counters.

An indirect method of measuring frequency is by first measuring wavelength with the use of a slotted line. The standing wave pattern in a slotted waveguide consists of maxima and minima due to the combination of incident and reflected waves. The distance between two adjacent maxima or minima is $\lambda_g/2$ where λ_g is the guide wavelength. This can be converted to frequency using a standard conversion factor.

A more accurate means of measuring frequency is by means of a resonant cavity or wavemeter. These are reaction type tunable cavities and when the cavity is tuned to resonance, a noticeable dip occurs in the transmitted power level of about 1 dB. This dip is observed on a meter or oscilloscope display, of the detected RF level. Frequency is read directly from the calibrated dial mechanism or is obtained from a calibration chart. The accuracy of cavity wavemeters depends upon the cavity Q, the tuning-dial calibration, mechanical backlash and on the effects of temperature and humidity. Cavity wavemeters have a high Q and give excellent accuracies between 0·5% and 0·005% if temperature-compensated by special construction.

Higher accuracy frequency measurements can be made with the use of a digital frequency counter to measure the result of two heterodyned signals. The standard signal used is generated harmonically from a crystal controlled oscillator and is mixed with the microwave signal. The output frequency is measured by a suitable frequency counter and it gives the microwave frequency directly as a digital display. The accuracy of the measurement is determined by the crystal stability which may be as good as 1 part in 10^{10} and on the accuracy of the frequency counter which has an error of ± 1 count.

Impedance measurements

At microwave frequencies, impedance is difficult to measure directly and is not employed very often. Moreover, its value in general depends upon the particular point in the waveguide at which the measurement is made. Since the impedance impedes the flow of energy, it can be measured by considering the transmission characteristics of the propagated energy.

Reflections due to impedance mismatches set up a standing wave pattern in the waveguide which can be measured or, alternatively, the incident and reflected powers can be compared. Consequently, the standing wave ratio or reflection coefficient are the parameters usually measured and from either of them, the impedance value can be obtained.

The measurement of the voltage standing wave ratio (VSWR) is accomplished with the use of a slotted section of waveguide wherein the pattern of electrical energy can be probed. When the effective impedance of a waveguide matches that of the source, no reflections occur and the VSWR is effectively unity. If an impedance mismatch is present, the incident and reflected waves combine vectorially to give a VSWR different from unity. This is known as the slotted waveguide method.

Measuring the intensity of the incident and reflected waves provides another means of obtaining the same result. This method requires an isolating device called a directional coupler with a high directivity for sampling the incident and reflected waves. The ratio of reflected to incident power is the reflection coefficient and can be measured by using the *reflectometer* method. However, the usual quantity which is measured is the *return loss* which is related to the reflection coefficient or the VSWR.

Slotted-waveguide method

(a) *Low VSWR*

The incident and reflected waves combine to produce maxima and minima as shown in Fig. 5.13. The VSWR has been defined as

$$\text{VSWR} = V_{max}/V_{min}$$

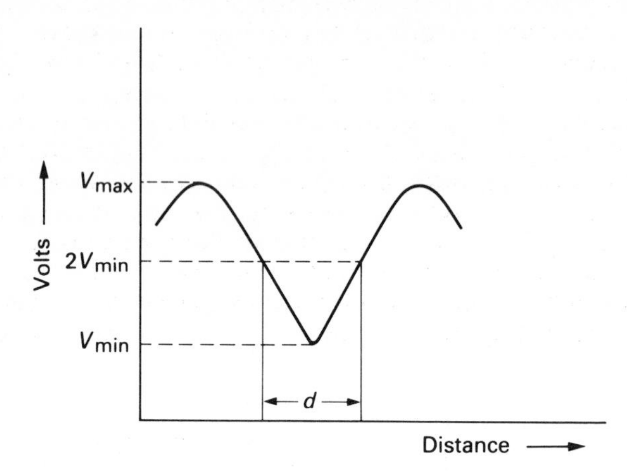

Fig. 5.13

The VSWR is measured with a slotted section of waveguide, as was illustrated in Fig. 5.12, which is terminated in a suitable mismatched load. A probe is loosely coupled to the slotted section and its output is fed into a high gain audio amplifier which is provided with a meter to read VSWR directly. By setting the probe at a maximum, the meter is adjusted to read unity and the probe is then moved to an *adjacent* minimum. The reading of the minimum can be directly calibrated in VSWR as it is the reciprocal of the actual meter reading. Hence, $\text{VSWR} = 1/V_{min}$ and if $V_{min} = 0{\cdot}5$, for example, the $\text{VSWR} = 1/0{\cdot}5 = 2{\cdot}0$.

(b) *High VSWR*
For a VSWR > 10, the square law of the detector crystal in the probe is no longer true and the meter calibration would not be correct. In this case, the probe is set to a minimum on the scale and its position is noted. It is then moved to two other positions (one on either side of the minimum) where the meter reading is twice the minimum value. If d is the distance between these two positions as is illustrated in Fig. 5.13, the VSWR is given by $\text{VSWR} = \lambda_g/\pi d$, where λ_g is the guide wavelength.

Reflectometer method[30]
Reflectometers measure reflection coefficient ρ or return loss (dB). The reflection coefficient ρ is a linear quantity varying between 0 and 1·0 and the return loss is defined as

$$\text{Return loss} = -20 \log_{10} |\rho| \text{dB}$$

which may vary from 0 when $|\rho| = 1{\cdot}0$ to ∞ when $|\rho| = 0$.

A reflectometer measurement can be made by using two directional couplers placed back to back. The forward coupler samples the incident signal from the source which is usually a sweeper and monitors its level on the forward detector. The reverse coupler samples the reflected signal from the device under test and senses it at the reverse detector.

A typical reflectometer measurement set-up is shown in Fig. 5.14. It consists of a sweeper such as a backward wave oscillator (BWO) which is a form of TWT that relies on the principle of velocity modulation. The unique characteristic of the BWO tube allows its frequency of operation to be uniformly changed with the application of a d.c. potential to the helix element within the tube. However, the forward power, must be levelled using a detector and the sweeper's automatic level control (ALC). Levelling, in addition to improving frequency response, also provides a better source match.

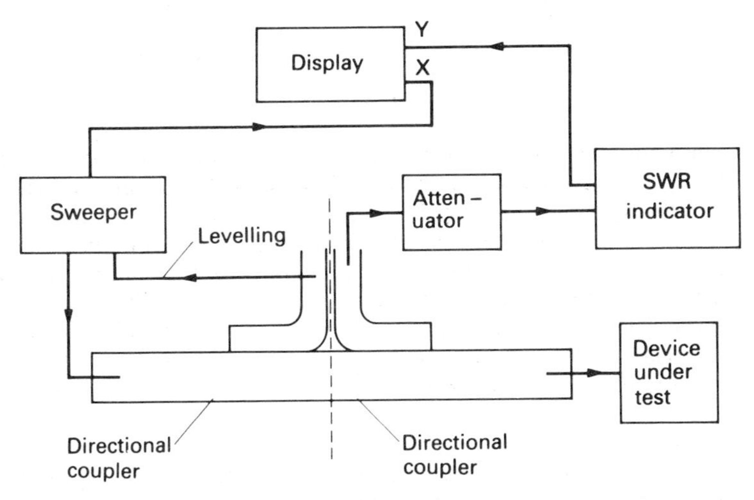

Fig. 5.14

The heart of the measurement centres around the dual three-arm directional couplers which are placed back to back. The most significant accuracy factor that contributes to the reflectometer method is the *directivity* of the reverse directional coupler. It must allow only an insignificant amount of forward energy to appear in the reverse direction. Typically, a 40 dB coupler directivity can be achieved which amounts to an error of only 40 dB below that of the unity reflected signal.

The reflectometer requires the establishment of a signal level based on a load element reflecting 100% of the energy i.e. a waveguide short. Since measurements by this method are to be made on devices under test which reflect less than 100% of the energy, a reference level is established with the short, then the short is replaced by the device under test.

In practice, much greater accuracy is achieved with the use of a precision variable attenuator which is placed ahead of the reverse detector, to reduce a unity reflected signal to an amount that equals the value of a lesser reflection due to the device under test. This value of attenuation in dB is the return loss of the device under test.

Power measurements

The measurement of power in a microwave circuit is fundamental. Unlike voltage or current, it is independent of the position in a waveguide and can be measured directly and conveniently by using thermally sensitive elements.

Microwave power measurements are either at low, medium or high power levels. Low power levels (< 10 mW) are measured by means of crystal rectifiers, bolometers, thermistors and, recently, by thermocouples. Medium power levels (< 1 watt) are measured by using calibrated attenuators with low power meters. However, thermocouples are also presently being used for direct power measurements above 100 mW. High powers (> 1 watt) are measured by using directional couplers and attenuators with low power meters or calorimeters.

A bolometer is a device whose resistance varies in proportion to the incident RF power as a result of the conversion of RF energy into heat energy. Bolometers are made with either thin strands of platinum wire which have a positive temperature coefficient and are known as *barretters* or they are made with small beads of semiconducting material with a negative temperature coefficient and are known as *thermistors*. Thermocouples, on the other hand, are devices that generate a d.c. voltage proportional to the RF power.

In the measurement of microwave power by a bolometer, the element is mounted in a short section of a waveguide. The bolometer is connected as one arm of a power bridge which indicates power by measuring its resistance. The more accurate bridges usually maintain the bolometer resistance constant (about 100–200 ohms) by substituting a.c. or d.c. power for the RF power.

For absolute power measurements, a thermal element must be properly mounted in the short section of waveguide so as to absorb all the incident power. Efficiencies around 95% are possible. To eliminate errors, mismatches should be tuned out as an error of 0·1 dB is possible with a VSWR of 1·3.

The use of thermocouples as sensing elements is a relatively recent development for measuring microwave power. Recent thermocouple designs using small, heat-sensitive elements have significant advantages such as low-drift, low VSWR and wide dynamic range, so that they are quickly displacing the bolometer.

A simplified block diagram of a thermocouple power meter is shown in Fig. 5.15.

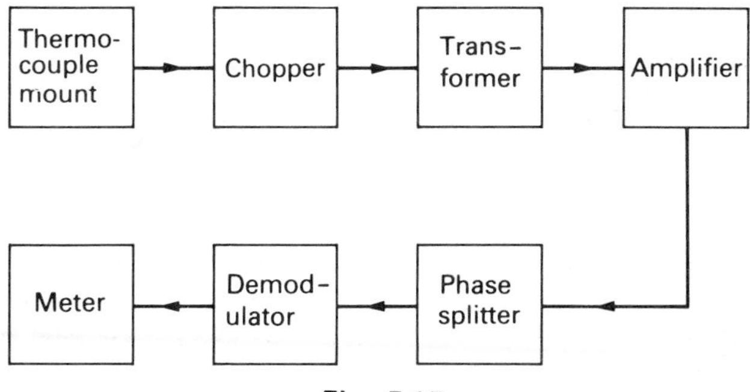

Fig. 5.15

In operation, the d.c. output voltage of the thermocouple element is applied to the input transformer and is modulated at a frequency of approximately 50–100 Hz, thus converting it from a d.c. voltage to an a.c. voltage. Modulation is accomplished through the use of a mechanical chopper which actually reverses the polarity of the input transformer. The a.c. signal is then amplified and applied to a phase splitter and a demodulator. The output of the demodulator is then applied to a d.c. meter.

With thermocouple power meters, accuracy is dependent on the gain of the instrument being matched to the power sensor's sensitivity. Thermocouple sensitivity is subject to change due to variations in temperature, overload and ageing. Hence, a convenient means for calibrating the instrument is vital. As an example, a power meter can normally provide an accurate, built-in 50 MHz power reference source for use in calibrating the meter and sensor combination.

Example 5.2

Give the sequence of the component units of a waveguide test-bench suitable for an accurate measurement of the voltage standing-wave ratio in a section of waveguide operating at approximately 3 GHz. Describe the procedure and state the precautions necessary to avoid errors.

A standing-wave indicator is terminated in an unknown load and has a detector with a square-law response. If movement of the probe produces deflections which vary between 35 and 10 % of full scale, determine the reflection coefficient of the load.

Figure 5.16 shows a typical set-up for measuring VSWR. The signal source is supplied with power from a stabilised power supply and is followed by an isolator which prevents any reflected power from reaching the source and thus ensures its stable operation. It is followed by a wavemeter for checking the frequency of the source and by an attenuator, which sets the level of power in the waveguide. The directional coupler then divides the power into two branches, one part being fed into a side-arm connected to a power meter, for monitoring the power output of the

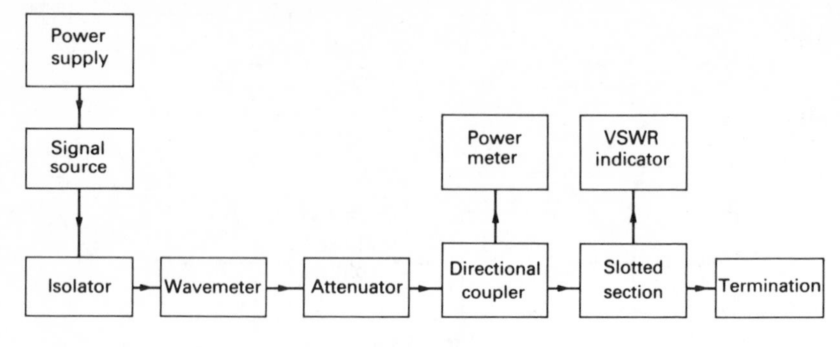

Fig. 5.16

source, while the remainder passes down the slotted section to the termination, which is either a short-circuit, an open-circuit or a matched load. The VSWR indicator is connected to the output probe of the slotted section.

Procedure

The source supply is switched on and the supply voltage is set to give a maximum output with 1 kHz square-wave modulation and the attenuator is adjusted to give a suitable deflection on the power meter. During the measurements, the power level on the power meter should be kept constant.

The wavemeter is now tuned to check the source frequency at 3 GHz. At resonance, a drop in power level is indicated on the power meter and by tuning the source mechanically, the point of *minimum* dip will give the correct frequency required. The wavemeter is then completely detuned so as not to interact with the rest of the circuit.

The probe carriage over the slotted section is now moved to and fro and the VSWR indicator will show that maxima and minima exist over the slotted section. To measure the VSWR, the method used depends on whether a low or high VSWR is present. Details were given in Section 5.3.

To avoid errors in the evaluation of the VSWR, the most sensitive scale should be used when measuring a low VSWR. When measuring a high VSWR, the position of a minimum is best determined by noting two positions on either side of it which give equal readings and the mean of these is the minimum reading. Furthermore, several readings should be taken of each measurement and the average obtained for the most accurate results.

Problem

Since the crystal detector used in the probe is a square-law device, its output current is proportional to the *square* of the voltage in the waveguide and hence

$$\text{VSWR} = \frac{V_{max}}{V_{min}} = \sqrt{\frac{I_{max}}{I_{min}}} = \sqrt{\frac{35}{10}}$$

or VSWR = 1·87.

Also $$\text{VSWR} = \frac{1+|\rho|}{1-|\rho|}$$

with $$|\rho| = \frac{\text{VSWR}-1}{\text{VSWR}+1} = \frac{0{\cdot}87}{2{\cdot}87}$$

or $$|\rho| = 0{\cdot}30$$

6

Optical communications

The transmission of optical frequencies through dielectric materials has been known for a considerable period of time, but in recent years only, it has become technically possible to develop the idea into a practical communication system by the use of glass fibres. Two important events which were responsible for this development were the advent of the laser in 1960 and the reduction of fibre losses to about 20 dB/km in 1970. The main components of such an optical fibre communication system are the optical source (transmitter), the glass fibre cable (transmission line) and a suitable detector (receiver).

Present technical developments have been very considerable because of the advances made in optical sources and low-loss cable manufacture. These developments are also due to the considerable advantages offered by such systems which include a large bandwidth capability, small size, low crosstalk, increased repeater spacing, greater reliability and a high immunity to outside electrical interference. Present day applications now extend from short-haul data link systems to high capacity long-haul telecommunication trunk systems.

6.1 Sources[31, 32]

The two main sources of light energy are the light-emitting diode (LED) and the solid-state laser. These devices are capable of providing sufficient power over the wavelength range 0·8 μm to beyond 1·3 μm, with sufficient reliability and at low-cost.

The LED is a *p-n* junction diode which, when heavily forward-biased, can be made to emit visible light. It is due to the recombination of electrons and holes when conduction band electrons are captured by valence band holes. The LED is characterised by the emission of spontaneous, incoherent light which is generated in a thin InGaAsP layer and it radiates in all directions. Heterojunctions are used to control the width of the recombination area.

In the double heterostructure shown in Fig. 6.1(a), electrons are injected under forward bias into the *p*-type narrow band-gap active region. The wider band-gap layers which sandwich the active region give efficient confinement of electrons and holes. The light power of about 1 mW can be coupled into an optical fibre with a loss of 15–20 dB which amounts to an output of about 20 μW.

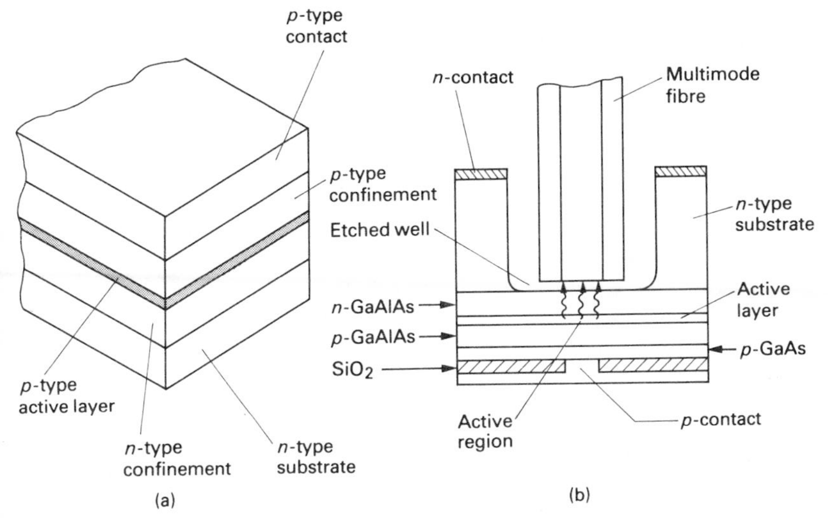

Fig. 6.1

The light power output of the LED increases linearly with the drive current and higher radiance LEDs (super luminescent diodes) are also available with coupled power outputs of about 200–400 μW. These diodes are either of the surface-emitting type or the edge-emitting type. The surface-emitting or Burrus diode which achieves high radiance by restricting the emission to a small area within a larger chip by using a small *p*-type contact. Light is taken out through a well, etched in *n*-type substrate and is shown in Fig. 6.1(b).

In contrast, the edge-emitting LED (ELED) makes use of light emitted in the junction plane increased by the guiding effect of the DH structure and a stripe contact is employed. There is no 'threshold' effect in LEDs and the LED can be amplitude-modulated with transmission rates of about 300 Mbit/s. A typical LED spectral response is centred on 900 nm and it has a spectral width of about 30 nm.

The laser diode or injection laser as it is often called, is a forward biased *p-n* junction diode in which electrons and holes recombine with the stimulated emission of coherent light. However, the coherent emission cannot begin until a minimum *threshold* current is reached which is typically about 150 mA. The emission bandwidth is about 2 nm and for pump currents of 10% to about 20% above threshold, the laser diode has a light output power of about 5–10 mW. The threshold effect produces a non-linear characteristic as shown in Fig. 6.2(a) and so the device is more suitable for pulse modulation techniques with possible bit rates of 1–2 Gbit/s.

Laser diodes developed for the 1·3 μm wavelength systems are made from an InGaAsP alloy grown on an InP substrate. A basic structure used is the

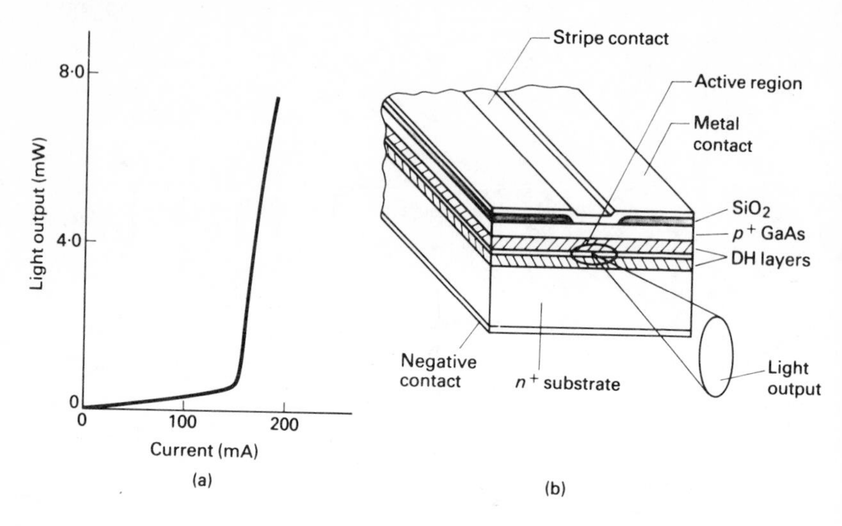

Fig. 6.2

stripe geometry DH configuration laser shown in Fig. 6.2(b). It consists of a very narrow InGaAsP layer of about 0·2 μm thick which is sandwiched between n and p-type InP layers. Optical emission occurs over a region defined by the contact stripe which is normally 5–20 μm wide.

Light is amplified in the plane of the active layer and is guided by the slab waveguide geometry. However, the laser is a more complicated device than the LED, relying on optical feedback from the cleaved semiconductor facets and having an active junction exposed at a free surface. Protective coatings of alumina or silica are generally used for the facets. Owing to the small dimensions of the emitting region, light can be coupled with high efficiency into a suitable fibre cable with a coupling loss of about 3 dB.

The differences in performance between the three optical sources can be illustrated in terms of their modulation properties, spectral response and output power capability. For analogue applications, the intensity of the optical power output of these devices may be varied by the AM signal and the technique is known as (AM-IM). If digital modulation is employed, all three devices may be used with a suitable binary code, because any non-linearity does not affect the amplitude of the output pulse.

The three optical sources are available with spectral outputs centred at specific wavelengths within a wide range of values and with different spectral linewidths. A wide range of optical powers is also obtainable with these devices and a typical set of performance parameters is summarised in Table 6.1.

Table 6.1

Device	*Material*	*Wavelength* (nm)	*Output power* (mW)	*Line width* (nm)	*Modulation type*	*Modulation bandwidth* (GHz)
LED	GaAs	900	0·5–0·8	40	Analogue	0·05–0·15
	InGaAsP	1300	0·1–0·5	100	or Digital (PCM)	0·05–0·15
ELED	GaAlAs	850	1–1·5	40	Analogue	0·1
	InGaAsP	1300	0·5–1·5	60	or Digital (PCM)	0·2
LASER	GaAlAs	850	1–3	3	Digital	0·5
	InGaAsP	1300	0·1–2	8	(PCM)	0·5–1·0

6.2 Fibre cables[33–35]

An optical glass fibre consists of an inner core which is surrounded by outer cladding material as shown in Fig. 6.3. The refractive index of the core is slightly higher than that of the surrounding material thereby confining the optical signal by total internal reflection. There are two types of optical fibres which are known as *monomode* fibres and *multimode* fibres. The 'modes' are the permitted field patterns of the optical signals which are able to propagate down the fibres.

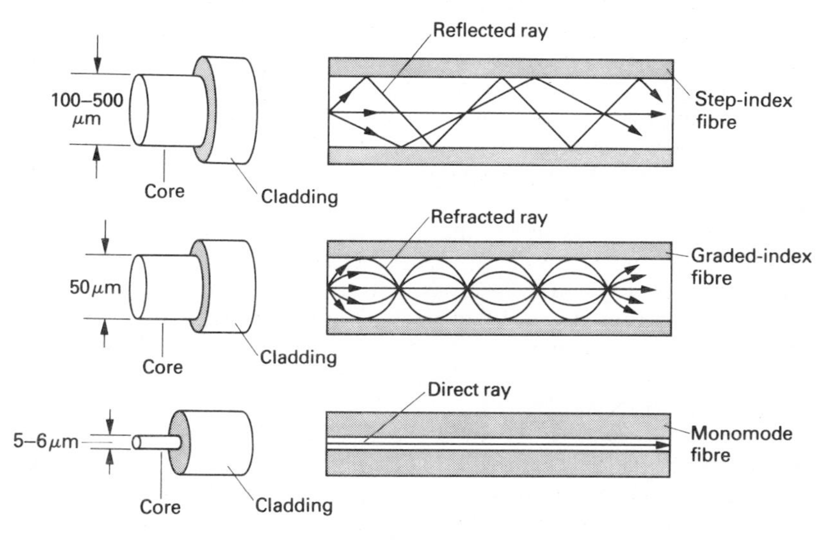

Fig. 6.3

Initially, monomode fibres were not generally available commercially because of their small core dimensions (typically a diameter of 5 μm), which limits manufacturing and practical application. Fewer difficulties in fabrication and application are met with, if fibres of larger diameter are used. Such fibres allow the propagation of not just one mode but of many modes. These multimode fibres may be of the step-index or graded-index form and are the most practical and common types used with core diameters between 20 μm and 150 μm.

Multimode fibres

In the step-index fibre, the modes propagate along a zig-zag path at greatly varying grazing angles so that some modes reach the end of the fibre by a longer route, while other modes do so by a shorter route. This leads to a large variety of mode transit times which limits the bandwidth of such fibres. These difficulties are less pronounced in graded-index multimode fibres which have a larger bandwidth because the transit times of the various modes are almost the same. Assuming the dimensions and range of refractive indices to be the same, there will only be half as many modes as in step-index fibres.

Another important parameter for evaluating optical fibres is the numerical aperture (NA) which is a measure of the degree of light acceptance of a fibre and is defined by

$$NA = n \sin \theta_{\mathrm{m}} = (n_1^2 - n_2^2)^{1/2} \simeq n_1 \sqrt{2\Delta}$$

where the refractive index $n = 1$ for air, θ_{m} is the maximum acceptance angle of the incident light, n_1 and n_2 are the refractive indices of the core and cladding material respectively, and $\Delta = (n_1 - n_2)/n_1$ is the relative refractive index difference. If the beam of light incident on the end face of the fibre is to enter the fibre, its incidence angle relative to the fibre axis must be smaller than the angle θ_{m}. Most communication fibres have numerical apertures lying within the interval $0{\cdot}14 < NA < 0{\cdot}22$ resulting in the interval $8° < \theta_{\mathrm{m}} < 13°$ for the maximum possible coupling angle and with light acceptance increasing as $(NA)^2$.

It is easiest to think about the fibre bandwidth in terms of the time domain rather than the frequency domain i.e. in terms of pulse spreading. In multimode fibres, we must distinguish between two mechanisms which give rise to such spreading, namely modal dispersion and material dispersion. Modal dispersion occurs in a step-index fibre because the energy of a pulse travels in several modes which have different group velocities and propagation paths through the optical fibre.

In the step-index fibre shown in Fig. 6.3 which has a well-defined core surrounded by cladding material of 1 % lower refractive index, rays incident at different angles θ follow different paths due to the various orders of modes, the lower-order modes following the shorter routes near the central axis and the higher-order modes following the longer off-axis routes. This produces pulse spreading or modal dispersion by an amount depending on the transit time difference between the shortest and longest routes.

By simple optical ray analysis, the modal dispersion effect gives rise to a group delay τ_d given by

$$\tau_d = t_{max} - t_{min} \simeq \frac{n_1}{c} L\Delta$$

where n_1 is the refractive index of the core, c is the velocity of light, L is the distance travelled along the fibre and Δ is the relative refractive index difference defined earlier. Since the pulse spreading increases in proportion with distance, a figure of merit of the information capacity of a fibre is given by the bandwidth-distance product $B \times L$. Typically, for a step-index fibre with a core refractive index of 1·5 and $\Delta \simeq 1\%$, $B \times L = c/n_1\Delta = 20$ MHz. km and $\tau_d \leqslant 50$ ns/km.

An effective method employed for reducing modal dispersion is by using graded-index fibre in which the refractive index profile in the radial direction r is parabolic and of the form

$$n(r) \simeq n_0 \left[1 - \Delta \left(\frac{r}{a} \right)^{\alpha} \right]$$

where n_0 is the refractive index of the core along the central axis, a is the core radius and the parameter $\alpha \simeq 2{\cdot}0$. Since the refractive index decreases, off-axis rays following the longer paths (higher-order modes) travel faster than those travelling near the central axis (lower-order modes) and this results in a quasi-equalisation of the transit times which minimises modal dispersion. The information capacity of a typical step-index fibre is thereby increased to about 1 GHz. km and $\tau_d \leqslant 1{\cdot}0$ ns/km.

The refractive index of glass varies with wavelength and material dispersion is caused by the variation of the refractive index. Since waves travelling in a multi-mode or step-index fibre are mainly confined to the core, the various signal frequency components in *each mode* travel with different phase velocities. Thus, the group velocity v_g of a light pulse varies since $v_g = c/N$ where c is the velocity of light, $N = \mathrm{d}(nk)/\mathrm{d}k$ is the group refractive index, $k = 2\pi/\lambda$ is the free-space propagation constant and n is the refractive index of glass.

The material dispersion effect causes a spreading of the light pulse as it travels along the fibre and the spreading is increased even further if the optical source has a large spectral width as in the case of an LED. The pulse spreading is proportional to the parameter $(\mathrm{d}^2n/\mathrm{d}\lambda^2)$ and the material dispersion effect is given by

$$\tau_m \simeq -\left(\frac{\lambda}{c} \right) L\lambda \left(\frac{\mathrm{d}^2 n}{\mathrm{d}\lambda^2} \right) \frac{\delta\lambda}{\lambda}$$

where L is the length of fibre, c is the velocity of light and $\delta\lambda/\lambda$ is the relative spectral width of the transmitted light pulse which includes both the spectral width of the source and the modulation bandwidth of the light pulse.

For a silica glass fibre at a wavelength of 0·85 μm, the delay parameter $(\lambda/c)(d^2n/d\lambda^2)$ is typically about 100 ps/km.nm and it yields a $B \times L$ product of about 2 GHz.km in practice. Nevertheless, it is found that at longer wavelengths, the material dispersion effect decreases to zero for most silica based fibres in the region of 1·3 μm and this is the reason for studying longer wavelength systems.

Multimode fibres are usually fabricated by the double-crucible process using conventional glass-making techniques starting with solid raw materials. When the two pure bulk glasses for core and cladding respectively have been prepared they are drawn into long rods. The core and cladding rods are then fed at a carefully controlled rate into the inner and outer regions of a suitably-heated concentric crucible arrangement from which fibres are drawn in the usual way. The process is illustrated in Fig. 6.4.

The double-crucible process lends itself well to large-scale manufacture, resulting in a low cost fibre of a highly reproducible quality. The refractive index profile can be graded and a wide range of core and cladding diameters and numerical apertures is possible. Losses of graded-index fibre with a

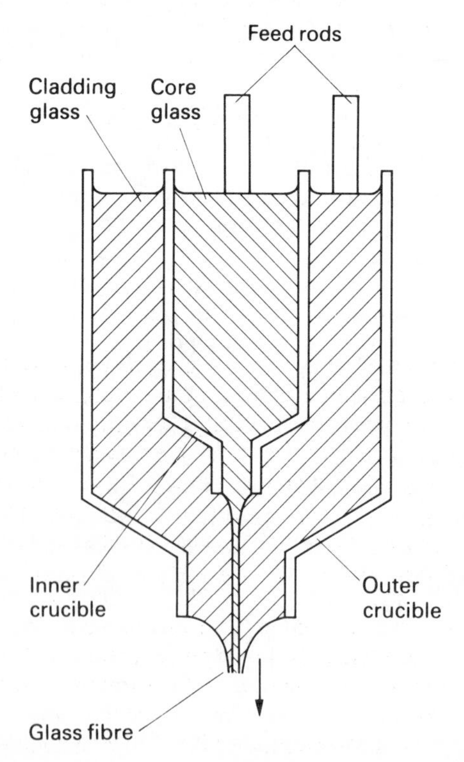

Fig. 6.4

numerical aperture of 0·2 lie in the range 3·5 to 5·5 dB/km and bandwidths range from 100 to 900 MHz.km depending on the glass composition. Fibres with larger numerical apertures (0·4 to 0·6) typically have losses of 10 dB/km.

Monomode fibres

Over recent years, the development of monomode fibre systems has progressed rapidly. The splicing and optical coupling problems have now largely been overcome and practical monomode systems that can exploit the extremely large bandwidth potential are soon becoming a reality.

In Appendix D, it is shown that the number of modes propagating in a multimode fibre depend on the normalised frequency V. By reducing the core radius such that $V < 2{\cdot}4$, only one mode (the fundamental mode) can propagate and a typical monomode fibre has a core diameter of about 5 μm. As modal dispersion is now absent, the two main dispersion effects are material dispersion (as in a multimode fibre) and waveguide dispersion.

Waveguide dispersion is due to the variation of the normalised propagation constant b which is a function of the core radius a and the wavelength λ i.e. it depends upon the ratio a/λ. This dependence is expressed by the dispersion parameter $V\mathrm{d}^2(bV)/\mathrm{d}V^2$ and the corresponding pulse dispersion effect is given by

$$\tau_{\mathrm{w}} \simeq \frac{L}{c}\Delta n_1 \frac{V\mathrm{d}^2(Vb)}{\mathrm{d}V^2}\left(\frac{\delta\lambda}{\lambda}\right)$$

with

$$V = \frac{2\pi a}{\lambda}(n_1^2 - n_2^2)^{1/2}$$

and

$$b = 1 - \frac{a^2(k^2 n_1^2 - \beta^2)}{V^2}$$

where L is the length of fibre, c is the velocity of light, $\delta\lambda$ is the spectral width, n_1 and n_2 are the refractive indices of the core and cladding material respectively, a is the core radius, β is the propagation constant and $k = 2\pi/\lambda$.

As material dispersion decreases at longer wavelengths and is of opposite sign to that of waveguide dispersion, the combined effect can be reduced to zero anywhere within the range 1·3 μm to 1·6 μm. This result increases the $B \times L$ product of a monomode fibre considerably and it is only limited by the birefringence effect arising from the ellipticity or stress in the core. This birefringence effect makes the two orthogonally polarised modes of the single-mode fibre distinguishable. The orthogonal modes have different propagation constants which produce pulse spreading. Normally, the pulse spreading may be as low as 1 ps/km and it amounts to a $B \times L$ product of around 1 THz.km.

Several methods of fabricating silica fibres have been developed and most of these use vapour phase reactions. The most widely used process is known as the modified chemical vapour deposition method (MCVD) which is illustrated in Fig. 6.5. The technique consists of two basic processes which are preform making and then drawing into a fibre. In the first stage, halide vapour materials are carried by oxygen into a uniform silica tube with good circularity to serve as a supporting structure.

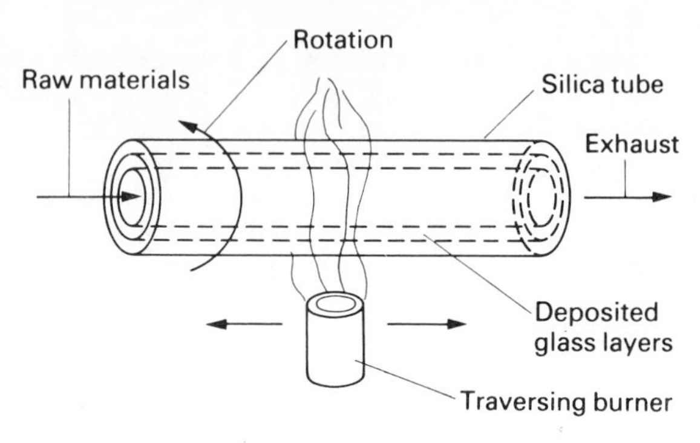

(a) Deposition process

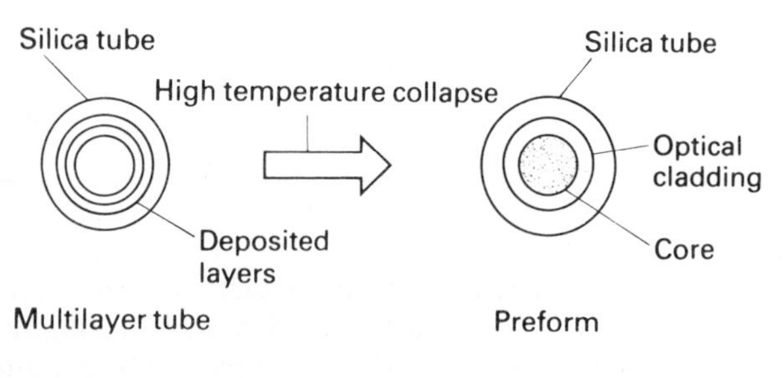

(b) Collapse process

Fig. 6.5

The tube is heated externally and the oxidation deposits a clean glassy layer on the inside of the tube. After depositing the cladding layer, the core layer is deposited and by doping the latter with germanium, the required refractive index difference is obtained. In the second stage, the composite tube formed is softened and made to collapse into a solid rod. This preform rod is mounted vertically in a fibre-drawing machine and a thin fibre is produced by uniform pulling. The fibre is then given a suitable plastic coating to protect it from surface abrasions.

Attenuation in an optical fibre is caused by various physical processes which give rise to absorption losses or scattering losses. A typical attenuation characteristic of monomode fibre is shown in Fig. 6.6. Its main

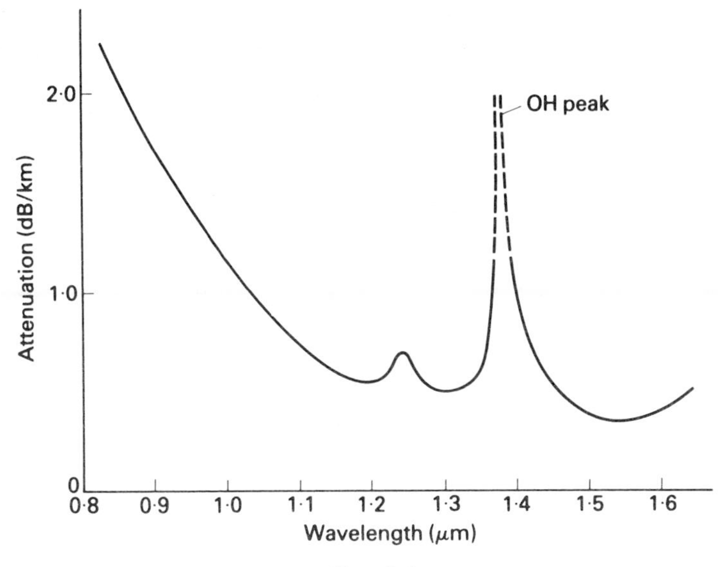

Fig. 6.6

features include two low-loss 'windows' at about 1·3 μm and 1·55 μm with values of attenuation around 0·5 dB/km and 0·3 dB/km respectively.

There is a scatter loss component due to Rayleigh scattering at the lower wavelength which varies as λ^{-4}. It is due to local variations in the refractive index of the material. The large absorption loss at about 1·4 μm is an OH vibrational loss associated with water contamination during the preform process. By improved manufacture, the various losses can be reduced to the ultimate limit of Rayleigh scattering at the shorter wavelengths and increasing only at the longer wavelengths because of infrared molecular absorption.

Optical fibre systems require the use of interconnection devices such as splices and connectors[36]. A splice is used for permanent jointing of cable while a connector provides a demountable facility. Fusion splicing using localised heating between two prealigned fibre ends, causing them to soften and fuse together, is commonly employed. Typical splicing losses are about 0·1 to 0·2 dB.

For mechanical coupling, a number of demountable connectors are employed such as the ceramic capillary connector and the triple ball connector. Coupling losses average between 0·5 dB and 1·0 dB approximately. The earlier techniques employed for multimode fibres are now being applied to monomode fibres but a much greater amount of control is required to achieve low losses consistently.

Example 6.1

A step-index fibre has an inner core of refractive index 1·5, a core diameter of 50 μm and a total length of 1·0 km. Light incident on the core makes a maximum refracted angle of 15° with the normal to the end face. What is the maximum bit rate which can be used with light pulses of 25 ns duration, if pulse dispersion is to be avoided?

Solution

In Fig. 6.7, light pulses travel along the direct path PQ and also along the indirect path RS. Due to the time delay between these paths, the pulses may overlap one another on arrival at the receiver and it gives rise to pulse dispersion. To determine this time delay, we have

$$d \sin 15^\circ = 50 \times 10^{-6}$$

or
$$d = \frac{50 \times 10^{-6}}{0{\cdot}2588} = 1{\cdot}932 \times 10^{-4}\ \text{m}$$

with
$$l = d \cos 15^\circ$$

or
$$l = 1{\cdot}932 \times 10^{-4} \times 0{\cdot}9659 = 1{\cdot}866 \times 10^{-4}\ \text{m}$$

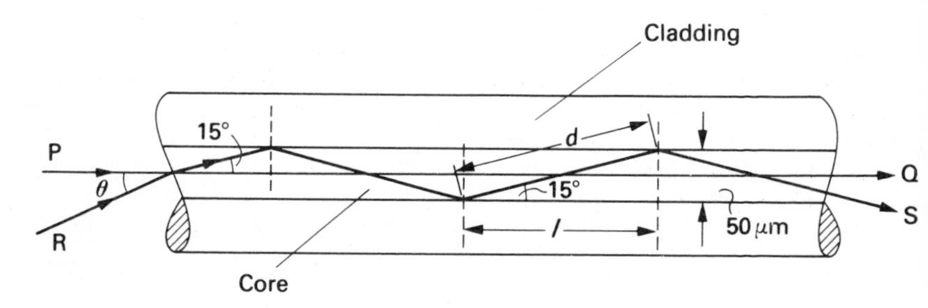

Fig. 6.7

The path length difference δ between the two rays PQ and RS is given by

$$\delta = \frac{(d-l)}{l} 1000 = \frac{(1{\cdot}932 - 1{\cdot}866)10^{-4} \times 1000}{1{\cdot}866 \times 10^{-4}}$$

or
$$\delta = 35\ \text{m}$$

If v is the velocity of propagation through the core, the corresponding time delay τ is

$$\tau = \delta/v = \frac{35}{(3 \times 10^8/1{\cdot}5)} = 175\ \text{ns}$$

For a bit rate B, the time interval T between pulses is $T = 1/B$ and to avoid pulse dispersion we must have

$$T \geqslant (\tau + 25 \times 10^{-9})$$

or
$$1/B \geqslant (175 + 25) \times 10^{-9}$$

Hence $$B \leqslant \frac{1}{200 \times 10^{-9}}$$

or $$B_{max} = 5{\cdot}0 \text{ Mbit/s}$$

6.3 Detectors and receivers[37, 38]

Photodetectors are used to convert optical energy into electrical energy at the receiving end of an optical fibre system. Semiconductor type photodiodes are best suited for this purpose because of sensitivity, size, speed of response, reliability and cost. The simplest photodiode is a *p-n* junction in a semiconductor with a band-gap energy less than the photon energy of the signal to be detected.

Photodetection occurs in the depletion region of the junction where the high electric field separates the electron-hole pairs which are produced when light is absorbed. The signal current appearing at the output is proportional to the light absorbed and the depletion region is maximised by reverse biasing the junction.

The performance of a simple *p-n* junction photodiode is greatly improved by a *p-i-n* structure (PIN diode) in which a lightly-doped intrinsic region is sandwiched between *p* and *n* regions as shown in Fig. 6.8. The device is normally operated with a reverse bias voltage which depletes the intrinsic region. This produces a high electric field, and photons falling on the carrier-void depletion region produce carriers with a near-maximum drift velocity which minimises response time. The device gives no gain and the signal requires further amplification.

For a digital system, a pulse must contain 10^4 photons if its arrival is to be detected against the thermal noise. Hence, to provide more internal gain,

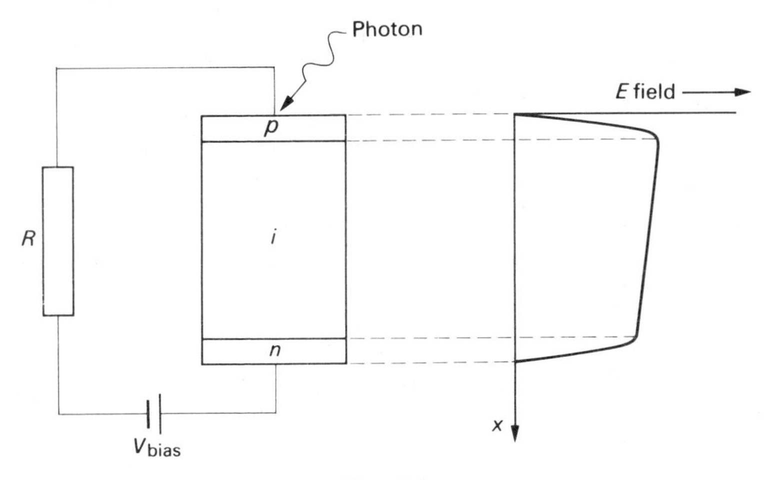

Fig. 6.8

avalanche multiplication may be used. The simplest form of device or APD using this technique is a *p-n* junction biased near to breakdown which causes impact ionisation or avalanche multiplication. During impact ionisation, a free electron or hole can gain sufficient energy to ionise a bound electron. The ionised carriers cause further ionisations, leading to an avalanche of carriers. The standard APD is a 'reach through' diode and is illustrated in Fig. 6.9.

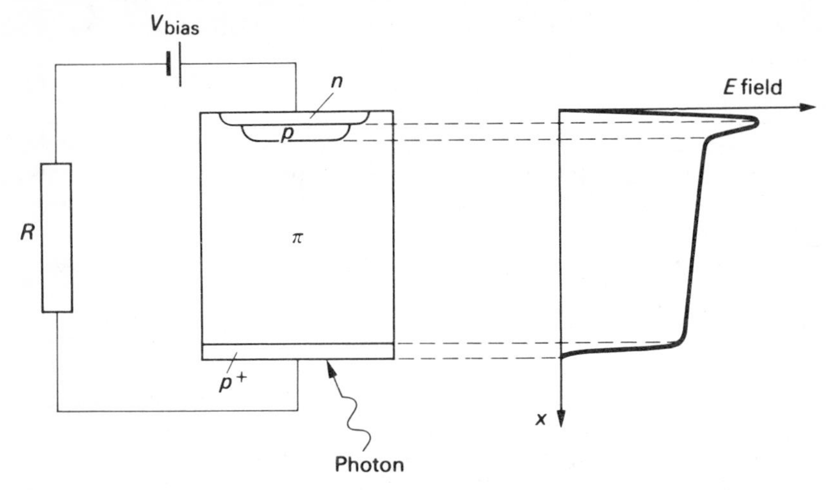

Fig. 6.9

The depletion region has a wide drift region and a narrow multiplying region as indicated in the electric field profile above. Photons are absorbed in the lightly doped 'π-region' whereas photogenerated carriers cause impact ionisation in the avalanche region. When the peak electric field is 5–10% below the avalanche breakdown field, the doping levels of the *p* and *n* avalanche regions allow the diode depletion region to 'reach through' to the low-doped π-region. Silicon APDs are suitable for the 0·8–0·9 μm region and germanium APDs at longer wavelengths. However, to overcome the problems associated with germanium APDs, much effort has been put into III–V alloy semiconductors, and in particular, InGaAsP devices.

A photodetector must have high performance characteristics, the foremost being a high sensitivity at the wavelength of operation, a minimum of output noise and a fast response speed. In addition, two important parameters of a photodiode are its *quantum efficiency* and *responsivity*. The quantum efficiency is a measure of the average number of electron-hole carrier pairs released by each incident photon and the responsivity is the ratio of the rms output current to the rms incident optical power. Typical performance parameters of some photodiodes are summarised in Table 6.2.

Table 6.2

Device type	*Material*	*Wavelength (nm)*	*Quantum efficiency (%)*	*Responsivity (A/W)*
APD	Si	800–900	70	0·7
	Ge	1300–1550	50	0·6
PIN	Si	800–900	70	0·65
	InGaAs	1300–1550	80	0·6

The signal output of the photodetector has to be amplified before demodulation and a typical optical receiver consists of a photodiode followed by a low-noise preamplifier. The receiver must have high efficiency, a fast response and a low level of noise in order to attain the desired signal-to-noise ratio or bit error rate for a given minimum amount of received optical power. Typical values for the S/N ratio of analogue systems are around 55 dB and for a digital system, the bit error rate (BER) is usually 1 in 10^9 which requires an *equivalent* S/N ratio of about 22 dB.

Two types of receiver design which may be used are the high impedance circuit and the transimpedance circuit which are illustrated in Fig. 6.10. In the first design, the photodiode load resistance is large and its thermal noise contribution is negligible. It is a simpler and cheaper circuit but needs a differentiating network to equalise the integrating effect due to the input capacitance. In the second design, the photodiode drives a lower input impedance feedback amplifier. The feedback resistor and amplifier open-loop gain are chosen so that little equalisation is required, yielding a

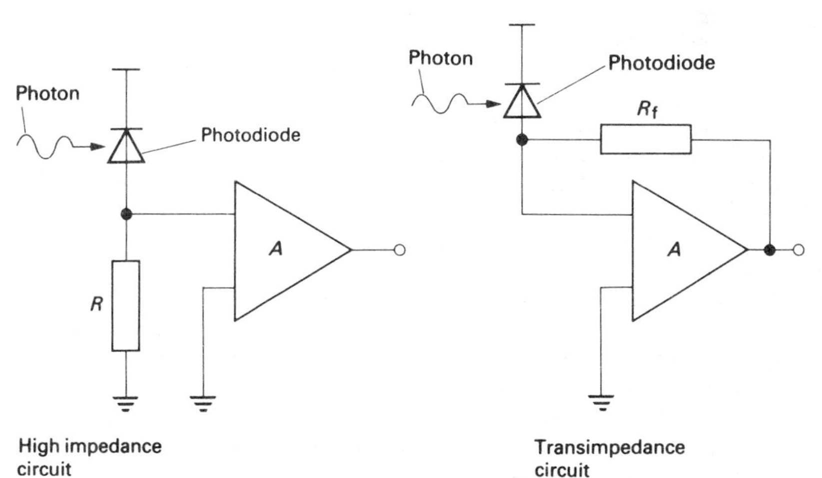

Fig. 6.10

high dynamic range. It is more suited to those systems where a high bandwidth (above 100 MHz) is required.

The principal figure of merit of a receiver is its sensitivity which is defined as the minimum optical power required (at a given data rate) for a specified signal-to-noise output ratio, in an analogue system or for a given output bit error rate, in a digital system. The ability of a receiver to achieve a specified performance level depends on the type of photodetector employed, the effects of noise in the system and on the characteristics of the various stages of amplification in the receiver. Typical performance parameters of some receivers used in systems are given in Table 6.3.

Table 6.3

Type	*Bit rate* (Mbaud)	*Sensitivity* (dBm)
APD (at 0·85 μm)	34	−54
	140	−48
	365	−41
PIN/FET (at 0·85 μm and 1·3 μm)	34	−48
	140	−42
	565	−34

A typical PIN-FET receiver employs the high impedance circuit and the photodiode is hybrid integrated with a low-noise amplifier using a GaAsMESFET as the front-end. Such receivers have a sensitivity of −44 dBm at 160 Mbaud and 40 dBm at 294 Mbaud when operating at 1·55 μm. The sensitivity can be improved further by reducing the input capacitance and increasing the FET transconductance. Small area photodiodes down to perhaps 30 μm diameter can be used with monomode fibre without undue alignment problems. Also, the gate length and gate width of the FET can be reduced from the present typical values of 0·7 by 300 μm. An improvement in receiver sensitivity of 7 dB or more seems possible, with performance beyond 1 Gbaud.

Example 6.2

In a step-index fibre, the refractive index of the core is 1·5 and that of the cladding 1·48. If the core diameter is 100 μm, calculate the value of the normalised frequency at wavelengths of (a) 0·8 μm and (b) 1·3 μm. How many modes propagate at each of these wavelengths? What core diameter will be required for single mode operation at 1·3 μm?

Solution

The normalised frequency V is given by the expression

$$V = \frac{\pi d}{\lambda}(n_1^2 - n_2^2)^{1/2}$$

where d is the core diameter, λ is the wavelength of the incident radiation, n_1 and n_2 are the refractive indices of core and cladding respectively.

(a) At $\lambda = 0{\cdot}8\ \mu m$ we have

$$V = \frac{\pi \times 100 \times 10^{-6}}{0{\cdot}8 \times 10^{-6}}[(1{\cdot}5)^2 - (1{\cdot}48)^2]^{1/2}$$

$$= \frac{\pi \times 10^3 \times 0{\cdot}245}{8}$$

or $$V = 96{\cdot}2$$

(b) At $\lambda = 1{\cdot}3\ \mu m$ we have

$$V = \frac{\pi \times 100 \times 10^{-6}}{1{\cdot}3 \times 10^{-6}}[(1{\cdot}5)^2 - (1{\cdot}48)^2]^{1/2}$$

$$= \frac{\pi \times 10^2 \times 0{\cdot}245}{1{\cdot}3}$$

or $$V = 59{\cdot}2$$

The number of modes propagating in each case is given by

$$N = V^2/2$$

Thus, at $\lambda = 0{\cdot}8\ \mu m$ we obtain

$$N = (96{\cdot}2)^2/2$$

or $$N = 4627 \text{ modes}$$

At $\lambda = 1{\cdot}3\ \mu m$ we obtain

$$N = (59{\cdot}2)^2/2$$

or $$N = 1752 \text{ modes}$$

For single mode operation at $1{\cdot}3\ \mu m$ we have the limiting condition

$$V = ua = 2{\cdot}405$$

where $u = \sqrt{k_1^2 - \beta^2}$, a is the core radius, $k_1 = n_1 k$, $\beta = n_2 k$ and $k = 2\pi/\lambda$. Hence

$$k_1 = 1{\cdot}5 \times 2\pi/1{\cdot}3 \times 10^{-6} = 7{\cdot}25 \times 10^6$$

$$\beta = 1{\cdot}48 \times 2\pi/1{\cdot}3 \times 10^{-6} = 7{\cdot}15 \times 10^6$$

and $$u = [(7{\cdot}25 \times 10^6)^2 - (7{\cdot}15 \times 10^6)^2]^{1/2}$$

$$= 10^6[(7{\cdot}25)^2 - (7{\cdot}15)^2]^{1/2}$$

or $$u = 1{\cdot}2 \times 10^6$$

Also

$$ua = 2{\cdot}405$$

with
$$a = \frac{2{\cdot}405}{1{\cdot}2 \times 10^{-6}} = 2{\cdot}0 \times 10^{-6}$$

or
$$2a = 4{\cdot}0\ \mu\text{m}$$

6.4 Further developments[39–41]

To meet the demands of many new telecommunication services such as high speed data systems, information technology and video transmission in the business and private fields, present developments taking place in monomode systems employing longer wavelengths will continue to expand. The enormous bandwidth capability of such systems will extend from the present position around 140 Mbit/s to well into the Gbit/s region.

Three notable developments likely to increase in importance in the late 1980s are wavelength division multiplexing (WDM), coherent detection and integrated optics. The present low attenuation achieved in the optical range suggests the possibility of sending down the same fibre several distinct wavelengths modulated by different information signals. This wavelength multiplexing technique which is illustrated in Fig. 6.11 can convey several signals in the same direction, thus leading to a potentially large increase in link capacity.

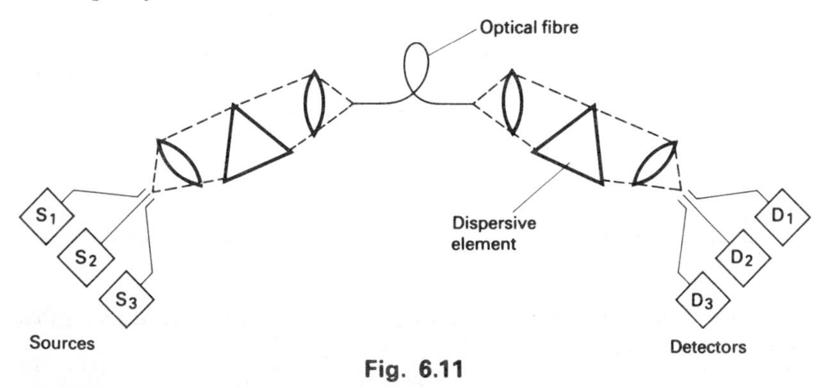

Fig. 6.11

Special optical elements have been designed to combine the various optical beams at the fibre input (multiplexers) and to direct them towards different photo-detectors (demultiplexers). The aim of such elements is to minimise insertion loss and crosstalk levels. A practical limit to the number of channels that can be simultaneously transmitted on a single fibre is set by the spectral width and the wavelength tolerance. The former restricts the number of LED channels to about three in the 1·3 μm low-loss window and to about five or six in the 1·6 μm low-loss window.

Up to now, optical fibre systems have tended to use direct intensity modulation of the optical source together with optical power detection at the receiver. More recently, coherent optical transmission and detection has been employed because it offers the possibility of much greater repeater spacing and ultimately, perhaps, dispensing with intermediate repeaters altogether.

A coherent optical transmission system employs a highly coherent and stabilised source which may be amplitude, frequency or phase modulated. At the receiver, a heterodyne technique is used by means of a local oscillator to detect the received signal coherently. Multiplication of the received signal with a strong local signal provides a considerable improvement in the output signal-to-noise ratio of the system, which sets the minimum detectable signal level close to that of quantum noise. Typical results obtained for a coherent system operating at 1·3 μm yielded a 15 dB signal-to-noise ratio improvement as compared with a non-coherent system, using a transmission rate of 565 Mbit/s over a distance of 62 km.

A further development in the area of optical waveguide components and circuits using planar technology is called *integrated optics*. Integrated optics involves the use of thin transparent dielectric layers on planar substrates as optical waveguides. One- or two-dimensional thin-film optical waveguides form the basis for integrated optical devices. A wide range of materials has been used for thin-film optical circuits, ranging from glasses and plastics to ferroelectric crystals and compound semiconductors. Typical optical devices using this technology include passive and active components such as directional couplers, modulators, switches and optical amplifiers. At present, a considerable amount of development work needs to be done to make integrated optical systems a practical reality in the future.

Example 6.3

The average power output of a photodetector is given by the expression

$$P = W \cdot B/2$$

where W is the incident optical energy in a pulse and B is the bit rate, assuming that equal numbers of zeros and ones are received.

For an error probability of 10^{-9} and a bit rate of 20 Mbit/s, determine the minimum optical power required in dBm at a wavelength of 900 nm. Assume that the bit error probability $P(\varepsilon) = \mathrm{e}^{-W/h\nu}$ where W has its previous meaning and $h\nu$ is the energy of a photon.

Solution

We have

$$P(\varepsilon) = \mathrm{e}^{-W/h\nu} = 10^{-9}$$

with $$W/h = 9 \log_e 10 = 20{\cdot}7$$

or $$W = 20{\cdot}7\, h\nu$$

Hence

$$P = 10{\cdot}35\, h\nu B$$

Substituting for $\lambda = 900$ nm and $B = 20$ Mbit/s we obtain

$$P = \frac{10{\cdot}35 \times 6{\cdot}63 \times 10^{-34} \times 3 \times 10^{8} \times 20 \times 10^{6}}{900 \times 10^{-9}}$$

or

$$P = 4{\cdot}57 \times 10^{-8}\ \text{mW}$$

with

$$P = 10 \log_{10}(4{\cdot}57) - 80\ \text{dBm}$$

or

$$P = -73\ \text{dBm}$$

Comment
An ideal photodetector has been assumed in the above example. In practice, for a photodetector with a quantum efficiency $\eta = 0{\cdot}8$, the optical power required is $P = 4{\cdot}57 \times 10^{-8}/0{\cdot}8\ \text{mW} = -72\ \text{dBm}$.

Problems

1. Sketch the form of the E and H fields, as appropriate, in
   (a) a coaxial line,
   (b) a rectangular waveguide operating in the mode with the lowest cut-off frequency,
   (c) a microstrip line.

   Give typical uses of each of these three types of interconnection method.

   Quoting typical parameters, give the order of attenuation, in dB/m, for a matched line using each of the types, at frequencies of 1 GHz and 10 GHz. Show the theoretical basis of the result for *one* of the cases cited. (C.E.I.)

2. A microstrip line of width w is supported on a ground plane by means of a dielectric substrate of relative permittivity ε_r and thickness h. Assuming that $w \gg h$ and neglecting any fringing effects, derive expressions for its characteristic impedance Z_0 and phase velocity v_p, using as a basis, transmission line theory.

   Obtain also, an expression for the instantaneous power, P_i, transmitted along the line, if the electric field intensity in the dielectric is E and the relative permittivity is 2·5.

3. Explain the causes and effects of attenuation and phase distortion in audio frequency cables.

   A cable is 32 km long and has the following distributed constants per loop km: $R = 25\ \Omega$, $L = 0{\cdot}6$ mH, $C = 0{\cdot}04\ \mu$F. Shunt conductance can be neglected. Calculate the value of the characteristic impedance at 10 kHz. If the cable is terminated with this impedance, calculate for a 10 kHz signal (a) the wavelength, and (b) the velocity of propagation.

   Calculate also, the gain in dB of a repeater inserted at the mid-point of the cable, to provide zero loss over the cable length. (U.L.)

4. At 4 MHz, a coaxial cable has the following distributed constants per loop km: $R = 68\ \Omega$, $L = 0{\cdot}28$ mH, $C = 0{\cdot}05\ \mu$F and $G = 0{\cdot}002$ S. Derive approximate expressions for the characteristic impedance and attenuation coefficient of the cable at this frequency.

   The cable is used with intermediate repeaters in a multichannel telephone system. If the attenuation at 4 MHz is not to exceed 50 dB between repeaters, calculate approximately the maximum repeater spacing allowable. (U.L.)

5. Explain what is meant by stub matching of high frequency transmission lines. What are the disadvantages of such a system?

A loss-free air-insulated transmission line, of characteristic impedance $50\,\Omega$, is connected to a resistive load of $120\,\Omega$. Derive expressions for (a) the position and (b) the length of a short-circuited stub that will terminate the main length of the line correctly. Hence determine the value of these quantities when the operating frequency is 85 MHz. The matching stub may be assumed to have a construction identical with that of the line, and the velocity can be taken as 3×10^8 m/s. (U.L.)

6 Determine the voltage standing wave ratio (VSWR) on a $75\,\Omega$ transmission line terminated by a resistance of $50\,\Omega$. Sketch the voltage standing wave pattern showing its phase relative to the position of the termination.

Give *two* reasons why the VSWR on the transmission line between transmitter and antenna should be minimised.

Use the Smith chart to design a single-stub matching system for a transmission line which is terminated by a normalised impedance of $0{\cdot}40 - j0{\cdot}10$. Use a short-circuited stub of the same characteristic impedance as the transmission line and connect the stub in parallel.

Show clearly on the Smith chart the stub length and stub position. (C.E.I.)

7 Starting from Maxwell's equations, obtain a solution for a plane electromagnetic wave of angular frequency ω in an infinite medium of absolute permeability μ, absolute permittivity ε and conductivity σ.

A plane electromagnetic wave of frequency 1 MHz passes normally into a large flat slab of copper ($\mu_r = \varepsilon_r = 1$, $\sigma = 58 \times 10^6$ S/m) of thickness much greater than the skin depth. The rms electric field strength of the transmitted wave at the copper surface is 1 V/m.

Determine (a) the average power flow per unit area at the slab surface, (b) the wavelength in the slab, and (c) the electric field strength at a depth of 20 μm. (C.E.I.)

8 A plane electromagnetic wave at a frequency of 3 GHz is travelling in air. The peak field intensity of the wave is 1 V/m and it is incident normally on a very large sheet of aluminium. Calculate the penetration depth of the wave in the aluminium sheet which has a conductivity $\sigma = 35 \times 10^6$ S/m and the power loss in the sheet per m^2 of surface area.

9 A rectangular waveguide having internal dimensions of $2{\cdot}0 \times 1{\cdot}0$ cm is energised in the TE_{10} mode at a frequency of 10 GHz.

(a) Using separate diagrams, sketch the *E* field and *H* field patterns. The broad and narrow dimensions of the waveguide and the field components must be clearly labelled.

(b) The waveguide is closed at one end and is to be strongly excited in this mode by a probe consisting of a short length of the inner conductor of a coaxial cable.

Explain precisely, with the aid of a diagram, where the probe should be inserted.

(c) Calculate the guide wavelength. (C.G.L.I.)

10 Discuss the propagation of a TE_{10} wave in a rectangular waveguide. What determines the lowest frequency which can be transmitted along the guide?

The output of an oscillator is fed into a coaxial line and also into a waveguide. The coaxial line has an air dielectric and losses are negligible. Probe measurements show that the distance between successive current nodes is 4 cm in the coaxial line and 5·4 cm in the waveguide.

Determine (a) the frequency of the oscillator, (b) the phase velocity and (c) the group velocity, in the waveguide. (U.L.)

11 A rectangular waveguide of internal dimensions 7·21 cm × 3·4 cm is operating in the TE_{10} mode at a frequency of 3 GHz and is delivering power to a matched load. If the peak voltage across the air-filled waveguide is 10 kV/cm, determine the maximum power delivered to the matched load.

12 A matched rectangular air-filled waveguide operating at 15 GHz supports only the TE_{10} mode. 15 GHz is 1·3 times the cut-off frequency of the TE_{10} mode and 0·7 times the cut-off frequency of the next higher-order mode. The peak value of the electric field strength in the guide is 1 kV/m. Determine (a) the guide dimensions, and (b) the average power transmitted down the guide. (C.E.I.)

13 Obtain an expression for the attenuation coefficient α of an air-filled copper waveguide propagating a TE_{10} mode in terms of the surface resistivity R_s, characteristic impedance Z_{TE}, free space wavelength λ and guide dimensions a, b.

For a copper waveguide with internal dimensions 4·75 cm × 2·1 cm and surface resistivity $1{\cdot}85 \times 10^{-2}$ ohm at a wavelength of 6 cm, determine the attenuation coefficient α for the TE_{10} mode in dB/m.

14 A cylindrical cavity of radius a and length l is operating at its resonant frequency ω_0 and supports a TM_{010} mode. If the Q of the cavity is given by the expression

$$Q = \omega_0 \times \left[\frac{\text{peak energy stored}}{\text{average energy loss}}\right]$$

obtain an expression for Q in terms of the dimensions of the cavity, the characteristic impedance Z_0 of free space and the surface resistivity R_s assuming air as the dielectric.

15 Explain, with the aid of an Applegate diagram, how a sinusoidal voltage coupled to the buncher cavity of a two-cavity klystron can result in an amplified signal at the output of the catcher cavity.

The d.c. accelerating voltage between cathode and buncher in a two-cavity klystron is 400 V. The voltage between the buncher electrodes has a frequency of 1 GHz and peak value of 80 V. Determine an appropriate distance between buncher and catcher.

Describe one application in communications in which the klystron is used in preference to solid-state devices. Explain the reason for this preference. (C.E.I.)

16 (a) Describe the principle of a method of impedance measurement which is based on the use of the slotted line and discuss the design factors that determine the accuracy of measurement.

(b) A 75 Ω solid-dielectric slotted coaxial line impedance measuring equipment is connected to an unknown impedance. At a frequency of 200 MHz, a VSWR of 1·9 and a voltage antinode 32·2 cm from the terminals appear on the measuring line. Assuming the dielectric to be loss-free and to have a relative permittivity of 2·5, calculate the value of the unknown impedance. (C.G.L.I.)

17 Describe an experimental method for the accurate determination of *large* standing-wave ratios which uses a moving-carriage standing-wave indicator with a square law detector. Measurements made in a slotted waveguide with such an instrument produced the following results.

θ	4	2	1	2	4
x	7·893	7·826	7·735	7·644	7·577

θ is the ratio of the observed output current of the square-law detector to that current as observed with the standing-wave indicator located at a position of a minimum deflection, and x is the reading of the vernier (in cm) which defines the position of the probe of the standing-wave indicator relative to a fixed reference plane.

Given that the guide wavelength was 4·00 cm, calculate the standing-wave ratio in the guide. (U.L.)

18 A section of waveguide terminated in an unknown impedance is connected to a waveguide test-bench. A VSWR of 4 is observed and minima are found when the detector probe is at 11·2 cm and 9·2 cm from the termination.

Mark on the Smith chart provided, points A and B corresponding to (a) the impedance of the load, and (b) the admittance of the load.

Determine the shortest distance from the load at which insertion of a purely reactive shunt matching device will reduce the VSWR to unity.

State with reasons whether the matching device should be inductive or capacitive. (C.G.L.I.)

19 (a) (i) Explain the meaning of the term optical waveguide.

(ii) Describe with the aid of a diagram how absorptive and radiative losses may occur in an optical fibre.

(b) State the law governing total internal reflection in any optical fibre waveguide.

(c) If the refractive index of an optical fibre waveguide core is 1·518 and the refractive index of the optical fibre cladding is 1·509, calculate the critical angle for a ray of light passing along the fibre so that transmission of the light is assured. (C.G.L.I.)

20 Determine the number of modes propagating at a wavelength of (a) 1·3 μm, and (b) 1·55 μm in a step-index fibre with a core diameter of 50 μm. The core is doped with germania to yield a value of $\Delta = 0{\cdot}012$ and the cladding is of pure silica with a refractive index of 1·465.

Compare the results obtained with that of a graded-index fibre whose parabolic profile parameter $\alpha = 2$.

21 The core and cladding material of an optical fibre cable have refractive indices of 1·48 and 1·46 respectively. The cable is excited for single-mode operation by a laser source of wavelength 1·35 μm. Calculate the maximum permissible core diameter in the following cases where α is the profile parameter

(a) step-index ($\alpha = \infty$)
(b) parabolic ($\alpha = 2$)
(c) triangular ($\alpha = 1$)

22 Derive an expression for the numerical aperture of a step-index fibre in terms of the refractive indices n_1, n_2 of the core and cladding material respectively, assuming meridonal rays only.

Evaluate the numerical aperture for an optical fibre with $n_1 = 1{\cdot}48$ and $n_2 = 1{\cdot}46$. What is the angle of the cone of light accepted by the fibre?

23 A multimode optical fibre link operates over a 5 MHz bandwidth and employs a receiver with a peak signal to rms noise ratio of 40 dB. An LED delivers optical power to the fibre and is intensity modulated with a modulation index of 0·5. If the detector is an APD with a noise factor of 10, a gain of 100 and a responsivity of 0·5 A/W, determine

(a) the minimum optical power required at the receiver,
(b) the shot noise current received.

Assume the amplifier noise to be negligible.

Answers

1 (a) trunk telephone line, 0·01 dB/m, 1·0 dB/m
(b) radar antenna feedline, 0·01 dB/m, 0·1 dB/m
(c) microwave integrated circuit, 0·1 dB/m, 1·0 dB/m.

2 $Z_0 = 377/\sqrt{\varepsilon_r} \times (h/w)$ ohm, $v_p = 3 \times 10^8/\sqrt{\varepsilon_r}$ m/s, $P_i = E^2 w^2/238$ watts

3 $\lambda = 19{\cdot}36$ km, $v = 19{\cdot}36 \times 10^4$ km/s, $\alpha = 28{\cdot}9$ dB

4 $l = 10{\cdot}5$ km

5 $l_1 = 0{\cdot}56$ m from the termination, $l_2 = 0{\cdot}47$ m

6 $l_1 = 0{\cdot}11$ from the termination, $l_2 = 0{\cdot}37$

7 (a) 2·65 mW/m², (b) $\lambda = 0{\cdot}42 \times 10^{-3}$ m, (c) $E = 0{\cdot}74$ V/m

8 $\delta = 1{\cdot}55 \times 10^{-6}$ m
Power loss = $2{\cdot}59 \times 10^{-7}$ W/m²

9 (b) The probe is inserted in the centre of the broad wall $\lambda_g/4$ from the closed end.
(c) $\lambda_g = 4{\cdot}54$ cm

10 $f = 1{\cdot}875$ GHz, $v_p = 4{\cdot}05 \times 10^8$ m/s, $v_g = 2{\cdot}22 \times 10^8$ m/s

11 1·17 megawatts

12 (a) $a = 1{\cdot}3$ cm, $b \leqslant 0{\cdot}7$ cm (b) 0·29 watts

13 $\alpha = 0{\cdot}034$ dB/m

14 $Q = \dfrac{1{\cdot}2Z_0/R_s}{(1 + a/l)}$

15 $d = 3{\cdot}47$ cm
Klystrons are used in UHF broadcast television which requires the transmission of power levels around 10 kW or more. Solid-state devices are unable to provide such high power levels at the present time.

16 (b) 50 − j30 Ω

17 VSWR = 7·0

18 $d = 0{\cdot}9$ cm
Shunt susceptance required is capacitive.

19 $\theta_c \simeq 83{\cdot}7°$

20 (a) 381, 193 (b) 268, 136

21 (a) 4·26 μm (b) 6·02 μm (c) 7·38 μm

22 NA = 0·242, Angle of cone = 28°

23 (a) $1{\cdot}28 \times 10^{-6}$ watts
(b) $I_{sh}^2 = 2{\cdot}05 \times 10^{-20}\ A^2/\mathrm{Hz}$.

References

1 RAMO, S. and WHINNERY, J. R. *Fields and Waves in Communication Electronics.* Chapter 6. John Wiley (1984).
2 SHEPHERD, J., MORTON, A. H. and SPENCE, L. F. *Higher Electrical Engineering.* Pitman (1970).
3 GIACOLETTO, L. J. *Electronics Designer's Handbook.* McGraw Hill (1977).
4 MORTON, A. H. *Advanced Electrical Engineering.* Pitman (1966).
5 BAHL, I. J. and TREVIDI, D. K. A Designer's Guide to Microstrip Line. *Microwaves,* May 1977.
6 BAHL, I. J. and GARG, R. A Designer's Guide to Stripline Circuits. *Microwaves,* January 1978.
7 YOUNG, L. (ed.) Advances in Microwaves, **8**, 67. Academic Press (1974).
8 EDWARDS, T. C. *Foundations for Microstrip Circuit Design.* John Wiley (1981).
9 *Reference Data for Radio Engineers.* Howard W. Sams Co. Inc. (1977).
10 SAAD, T. S. *Microwave Engineers Handbook,* Volume 1. Artech House (1971).
11 WOLF, H. F. *Handbook of Fibre Optics.* Granada (1979).
12 LACY, E. A. *Fibre Optics.* Prentice Hall (1982).
13 FREEMAN, R. L. *Transmission Handbook.* Chapter 14. John Wiley (1981).
14 CONNOR, F. R. *Signals.* Edward Arnold (1982).
15 SMITH, P. H. Transmission Line Calculator. *Electronics,* January 1939, and An Improved Transmission Line Calculator. *Electronics,* January 1944.
16 RAMO, S. and WHINNERY, J. R. *Fields and Waves in Communication Electronics.* Chapter 8. John Wiley (1984).
17 STRATTON, J. A. *Electromagnetic Theory.* McGraw Hill (1941).
18 JORDAN, E. C. and BALMAIN, K. G. *Electromagnetic Waves and Radiating Systems.* Prentice Hall (1968).
19 LORRAIN, P. and CORSON, D. *Electromagnetic Fields and Waves.* W. H. Freeman and Co. (1970).
20 SANDER, K. F. and REED, G. A. L. *Transmission and Propagation of Electromagnetic Waves.* Cambridge University Press (1978).
21 BADEN FULLER, A. J. *Microwaves.* Pergamon Press (1979).
22 REICH, H. J. (ed.) *Microwave Theory and Techniques.* D. Van Nostrand Co. N.Y. (1953).
23 LIAO, S. Y. *Microwave Devices and Circuits.* Prentice Hall (1980).
24 HOWES, M. J. and MORGAN, D. Y. (ed.) *Microwave Devices.* John Wiley (1976).
25 GUNN, J. B. *Solid State Communications* **1**, 88, 1963.
26 THOMAS, H. E. *Handbook of Microwave Techniques and Equipment.* Prentice Hall (1972).
27 NARDA MICROWAVE CORPORATION, Catalogue No. 22.
28 BAILEY, A. E. (ed.) *Microwave Measurement.* Peter Peregrinus (1985).
29 HOOK, A. P. Microwave Scalar Analysis. *Electronics and Power,* February 1983.

30 ADAM, S. F. *Microwave Theory and Applications.* Prentice Hall (1969).
31 NEWMAN *et al.* Sources for Optical Fibre Communications. *Telecommunication Journal*, **48**, 673, 1981.
32 GOWAR, J. *Optical Communication Systems.* John Wiley (1984).
33 GARDNER, W. B. Fundamental Characteristics of Optical Fibres, *Telecommunication Journal*, **48**, 638, 1981.
34 HOWES, M. J. and MORGAN, D. V. *Optical Fibre Communications.* John Wiley (1980).
35 Optical Fibre Communications. *The Radio and Electronic Engineer*, **51**, July/August 1981.
36 KEISER, G. *Optical Fibre Communications.* McGraw Hill (1983).
37 MILLER, J. Advances in Electron Physics, **55**, 189, 1981.
38 SMITH *et al.* Receivers for Optical Fibre Communication Systems. *Telecommunication Journal*, **48**, 680, 1981.
39 GAMBLING, W. A. The Development of Optical Communication. *Electronics and Power*, November/December 1983.
40 LILLY, C. J. The Application of Optical Fibres in the Trunk Network. *Telecommunication Journal*, **48**, 109, 1982.
41 SMITH, D. R. Advances in Optical Fibre Communications. *Physics Bulletin*, **33**, 401, 1982.
42 MOSIG, J. R. and GARDIOL, F. E. Advances in Electronics and Electron Physics, **59**, 1982/83.
43 SCHNEIDER, M. V. Microstrip Lines for MICs. *Bell System Technical Journal*, **48**, 1421–44, May/June 1969.
44 TURNER, L. W. (ed.) *Electronic Engineer's Reference Book*. Newnes-Butterworth (1976).
45 HARVEY, A. F. *Microwave Engineering*. Academic Press (1963).
46 SNYDER, A. W. Asymptotic Expressions for Eigenfunctions and Eigenvalues of a Dielectric Optical Waveguide. *IEEE Transactions on Microwave Theory and Techniques*, MTT 17, 1130–38, 1969.
47 GLOGE, D. Weakly Guiding Fibres. *Applied Optics*, **10**, 2252–8, October 1971.

Appendices

Appendix A: Stripline and microstrip[42,43]

At higher frequencies, the performance of stripline and microstrip is affected by moding effects due to higher-order modes, and, in the case of microstrip, also by dispersion. Thus, it is observed that the maximum frequency of operation in stripline is limited by the excitation of TE modes. For wide strip the cut-off frequency of the lowest order TE mode is given by

$$f_c = \frac{30}{\sqrt{\varepsilon_r}\,(2w + \pi h/2)}\ \text{GHz}$$

where w and h are in centimetres.

Furthermore, for the TE mode, the Q-value and cut-off frequency in stripline depend on the spacing between the ground planes i.e. on the parameter h, in addition to several other parameters. In general, a wider spacing on h will increase the Q-value, but it simultaneously decreases the cut-off frequency of the TE mode of operation.

The maximum frequency of operation in microstrip is also limited by the excitation of spurious modes in the form of surface waves and transverse resonances. The surface waves may be TM or TE modes which propagate between the dielectric and the ground plane. A frequency for strong coupling between the quasi-TEM mode and a TM mode occurs when the two modes have the same phase velocity which occurs at a frequency given by

$$f_{TEM} = \frac{c \tan^{-1} \varepsilon_r}{\pi h \sqrt{2(\varepsilon_r - 1)}}$$

For a wide microstrip line, a transverse-resonant mode can strongly couple to the quasi-TEM microstrip mode. The cut-off frequency for this mode is given by

$$f_c = \frac{c}{\sqrt{\varepsilon_r}\,(2w + 0{\cdot}8h)}$$

and the mode can be suppressed by inserting slots into the metal strip.

A further feature which occurs in microstrip lines is the change in the effective dielectric constant and characteristic impedance as frequency increases, thus making the microstrip line *dispersive*. The dispersion is due to the propagation of hybrid modes along the microstrip line.

The effect is accounted for by defining a frequency-dependent effective microstrip permittivity $\varepsilon_{eff}(f)$. At the low-frequency end, it is equal to the static-TEM value of ε_{eff} and as the frequency increases, it approaches the substrate value of ε_r. The frequency f_d below which dispersive effects may be neglected is given by

$$f_d = 0{\cdot}3\sqrt{\frac{Z_0}{h\sqrt{\varepsilon_r - 1}}}\ \text{GHz}$$

where h is in centimetres.

As in the case of waveguides, microwave measurements can be performed on microstrip using frequency domain or time-domain techniques. In the former case, swept frequency measurements may be employed to determine the magnitude of a reflection coefficient using a dual directional coupler, a coaxial-to-microstrip transition and a microstrip line as the device under test. Such an arrangement can also be used with other microstrip circuits for making dispersion or Q-value measurements.

The use of microstrip has led to the widespread development of microwave integrated circuits (MICs), since semiconductor devices can easily be integrated with the microstrip line, by use of a semiconducting material as the substrate. MICs are finding wide application at microwave frequencies both, for the manufacture of components such as couplers and circulators or for complete systems such as receivers.

Appendix B: Ferrites[44,45]

Certain ceramic materials with good magnetic properties, like ferromagnetic substances, and a high resistivity, like an insulator, are called *ferrites*. Typical examples of such ferrimagnetic materials are substances like XFe_2O_4 where X is a divalent metal like cobalt, nickel, magnesium etc. and also certain magnetic oxides of the garnet family such as yttrium-iron-garnet $Y_3Fe_5O_{12}$ (YIG). Unlike ferromagnetic materials, ferrites can interact with electromagnetic waves at microwave frequencies and they yield interesting non-reciprocal properties when they are also under the influence of an external d.c. magnetic field.

The magnetic properties of a ferrite are due to the presence of loosely bound or unpaired electrons. These spinning electrons behave as 'magnets' with their magnetic moments aligned along the spin axis. Under the influence of an external magnetic field, the electron magnets tend to line up in the direction of the field or in a position of minimum energy. If an R.F. field with the right polarisation is now applied at right angles to the external magnetic field, the electrons tend to precess about the d.c. magnetic field like a gyroscope and it is called the *gyromagnetic effect*.

When the precessional frequency of the electrons and the external magnetic field have the correct values, there is a transfer of energy from the R.F. field to the precessing electrons which gives rise to the phenomenon known as resonance absorption. This condition is given by

$$\omega = \gamma H_0$$

where ω is the Larmor frequency, γ is the gyromagnetic ratio and H_0 is the external d.c. magnetic field which is illustrated in Fig. B.1.

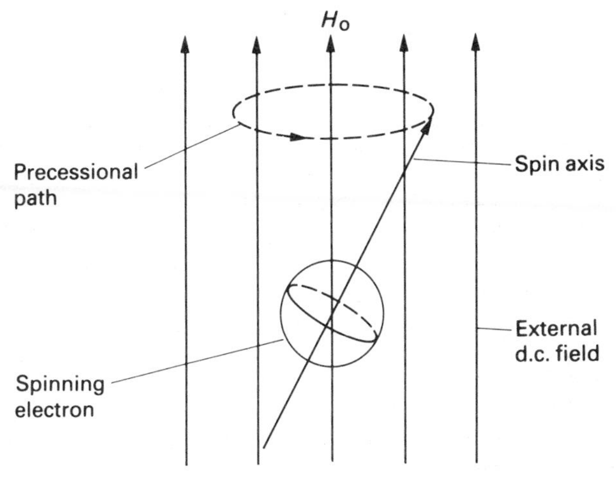

Fig. B.1

The gyromagnetic effect in ferrites illustrates that energy can be coupled between the ferrite electrons and the applied R.F. field, the coupled energy being dissipated as heat in the ferrite material. This interchange of energy is a function of the direction of propagation through the material and it depends on the polarisation of the R.F. wave. In one direction of polarisation, energy-loss is low, while in the opposite direction it is high, and so it accounts for the non-reciprocal transmission properties of the ferrite.

Some further parameters of interest in a ferrite are its saturation magnetisation, line width and Curie temperature. The line width is the range of magnetic field values over which resonance absorption takes place and the bandwidth of the ferrite is determined by its line width.

A narrow line width substance like YIG yields typical Q-values around 10 000. The Curie temperature is the well-known temperature at which the substance loses its magnetic properties. It sets an upper limit on the operating power level of the material and this in turn determines its maximum operating power level.

Ferrite devices are used extensively in radar and microwave communication systems. Two well-known devices are the isolator and the circulator. The isolator is a device for producing low loss in the forward direction but a large loss in the reverse direction. Thus, an incident wave in the forward direction is transmitted with little loss but a reflected wave travelling in the reverse direction is considerably attenuated. The device is able to isolate a microwave source from a mismatched load and so improve its frequency stability.

In a typical resonance absorption device which is shown in Fig. B.2, a slab of ferrite in the form of a rectangular bar is placed near one end of the rectangular guide about a quarter of the length of the broader wall since at this point, the R.F. magnetic field is strongly circularly polarised for the TE_{10} mode. On application of a d.c. magnetic field H_0, the electrons will precess about the H_0 field and absorb energy from the TE_{10} mode if its circular polarisation is in the right direction.

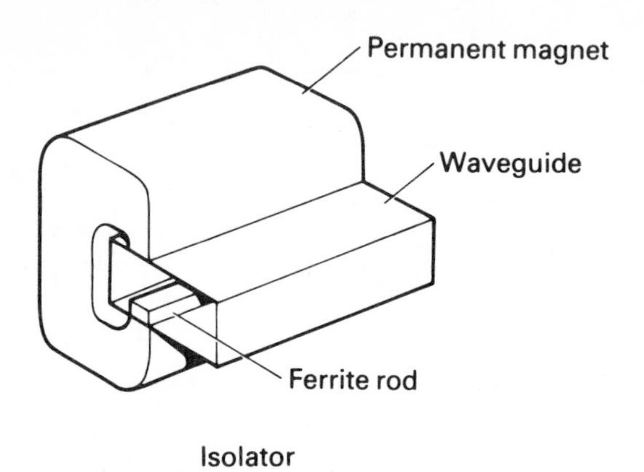

Isolator

Fig. B.2

The TE_{10} wave will be strongly absorbed in one direction, but not in the reverse direction, because the R.F. magnetic field of the returning TE_{10} is reversed in direction and this does not assist precession. A typical low-power device weighing about 2 kg has an average dissipation of 15 watts, an insertion loss of 1 dB and provides an isolation of better than 30 dB.

The other well-known device is the ferrite circulator which is used for coupling power in one direction only but not in the opposite direction. The device may be of the three-port or four-port design using stripline or waveguide construction and it is used for connecting a transmitter or receiver to the same antenna.

The three-port circulator illustrated in Fig. B.3 consists of three waveguide arms meeting at a junction. In the centre, is a ferrite post magnetised by a polarising d.c. field H_0. Power entering port 1 is made to move around to port 2 and then to port 3. Similarly, power entering port 2 will rotate to port 3 and then to port 1. This clockwise rotation is due to the wavefront of the wave being rotated by the ferrite material in the clockwise direction. Thus, port 3 can be isolated from port 1 if all the power is coupled out from port 2. This is the case when a transmitter is connected to port 1 and an antenna is connected to port 2.

By connecting two three-port devices together, a four-port device is obtained as the fifth port is usually terminated in a matched load, with the

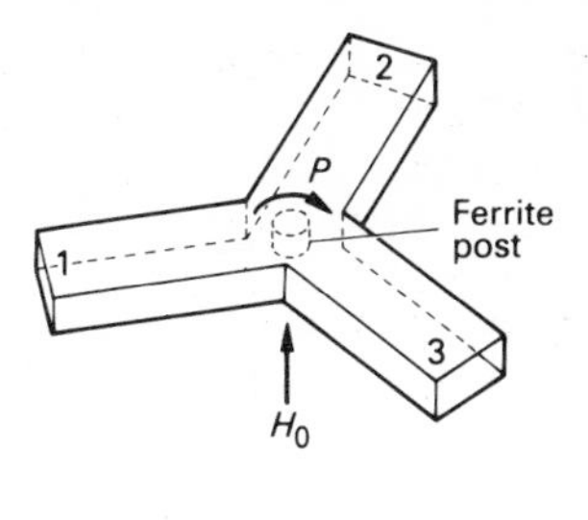

Circulator

Fig. B.3

fourth port connected to a receiver. A typical four-port stripline circulator may have an insertion loss of about 1 dB and provides about 40 dB isolation between transmitter and receiver.

Appendix C: Cavity resonators

Fields can exist in regions entirely bounded by conducting walls called cavities. The cavities may be rectangular or circular in shape and behave very much like resonant circuits. If a small loop is coupled into the cavity and excited by an oscillator, very large fields can exist in the cavity at certain frequencies only. The general principle is that standing waves are set up in the cavity which satisfy the boundary conditions at the specific frequencies, if the cavity dimensions are correct.

Q of a cavity

The Q of an inductance or capacitance is a measure of its ability to act as a pure reactance. There will be dielectric or ohmic losses that will cause a small energy loss every time energy is stored in the reactance. For a coil of inductance L in series with a resistance R, the Q is given

$$Q = \omega_0 L/R$$

where ω_0 is the resonant angular frequency

or
$$Q = \omega_0 \frac{\frac{1}{2}LI_{\mathrm{m}}^2}{\frac{1}{2}I_{\mathrm{m}}^2 R} = \omega_0 \frac{\text{(peak energy stored)}}{\text{(average energy loss)}}$$

This latter definition is more appropriate for cavities where $Q = Q_{\mathrm{U}}$, the *unloaded* Q of the cavity if the average energy loss occurs in the cavity only. Denoting peak stored energy as U and average loss as W, we have

$$U = \int_V \frac{\varepsilon E_{\mathrm{m}}^2}{2}\,\mathrm{d}v = \int_V \frac{\mu H_{\mathrm{m}}^2}{2}\,\mathrm{d}v$$

If there are no dielectric losses but only ohmic losses in the cavity walls, we have

$$W = \int_S \frac{|H_t|^2 R_s}{2} da$$

where $|H_t|$ is the tangential magnetic field component which numerically equals the surface current and R_s is the surface resistivity.

Now $$R_s = 1/\sigma\delta$$

where σ is the conductivity and δ is the penetration depth given by

$$\delta = \sqrt{\frac{2}{\omega_0 \mu \sigma}}$$

Hence $$Q_U = \frac{\omega_0 \mu \sigma \delta \int_V H^2 \, dv}{\int_S H^2 \, da} = 2/\delta \frac{\int_V H^2 dv}{\int_S H^2 \, da}$$

or $$Q_U = 2/\delta \frac{(\text{mean of } H^2 \text{ in the volume} \times \text{volume})}{(\text{mean of } H^2 \text{ over the surface} \times \text{surface})}$$

with $$Q_U \simeq 2/\delta \frac{\text{Volume}}{\text{Area}}$$

if the mean of H^2 in the volume and over the surface are approximately equal for the cavity.

If the cavity is coupled to a load as in Fig. C.1, the loaded Q is given by Q_L where

$$Q_L = \omega_0 \frac{(\text{stored energy})}{\text{energy loss in cavity \& load}}$$

Hence

$$\frac{1}{Q_L} = \frac{\text{energy loss in cavity}}{\omega_0 \times \text{stored energy}} + \frac{\text{energy loss in load}}{\omega_0 \times \text{stored energy}}$$

or $$\frac{1}{Q_L} = \frac{1}{Q_U} + \frac{1}{Q_{\text{Load}}}$$

which means that the loaded Q of the cavity is less than its unloaded value. Energy may be coupled into and out of the cavity by means of loops, as in Fig. C.1.

Cylindrical cavities are commonly used in practice as they have a large volume to surface area which gives a high value of Q_U.

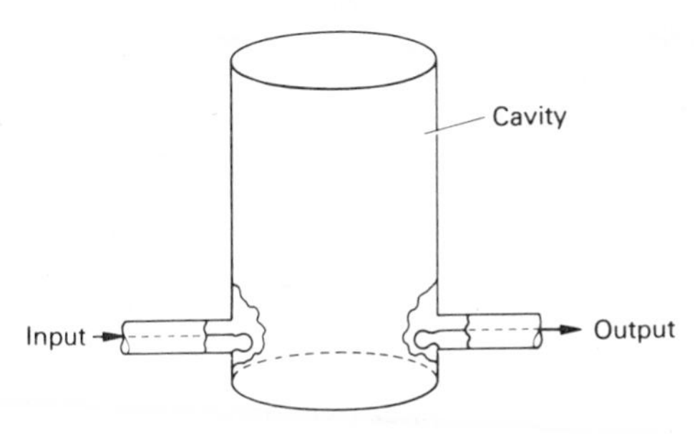

Fig. C.1

Cavity modes

These are basically waveguide modes since a resonant cavity can be constructed by shorting a length of waveguide (rectangular or circular) with metal end plates. A standing wave pattern will exist in the cavity provided the boundary conditions at the metal walls are satisfied. This is achieved if the cavity length l is an integral number of $\lambda_g/2$.

Cavity modes are specified by the two basic subscripts m, n used for rectangular waveguides. In addition, a third subscript p or q designates the number of half wavelengths along the cavity length l. Hence, we have $TE_{mnp}(H_{mnp})$ modes or $TM_{mnp}(E_{mnp})$ modes for rectangular cavities and $TE_{mnq}(H_{mnq})$ modes or $TM_{mnq}(E_{mnq})$ modes for circular cavities.

Rectangular cavities

Let the cavity dimensions be a, b, l as shown in Fig. C.2. The additional boundary condition to be satisfied is given by

$$l = p\lambda_g/2$$

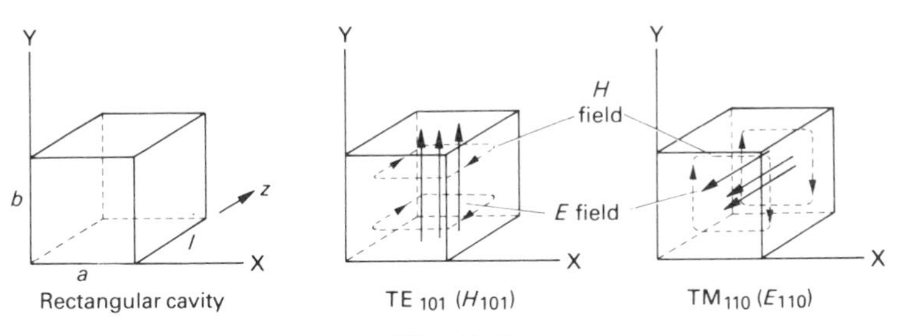

Fig. C.2

where $p = 0, 1, 2$ etc. and the condition $p = 0$ implies a mode in which the electric field lines are normal to the end walls, in the z-direction.

Since the cavity mode is basically a waveguide mode, we have

$$1/\lambda^2 = 1/\lambda_c^2 + 1/\lambda_g^2$$

with

$$\lambda_g = 2l/p$$

Since

$$\lambda_c = \frac{2}{\sqrt{(m/a)^2 + (n/b)^2}}$$

the resonant wavelength λ of the cavity is given by

$$1/\lambda^2 = \frac{(m/a)^2 + (n/b)^2}{4} + p^2/4l^2$$

or

$$\lambda = \frac{2}{\sqrt{(m/a)^2 + (n/b)^2 + (p/l)^2}}$$

for TE_{mnp} or TM_{mnp} modes.

Cylindrical cavities

These cavities are easier to construct mechanically with a high Q and are greatly used in practice. The cavity modes are designated TE_{mnq} or TM_{mnq} where q refers to the number of $\lambda_g/2$ along the length l indicated in Fig. C.3.

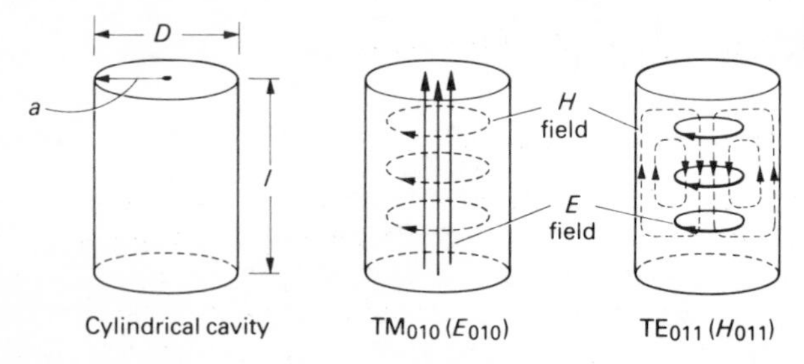

Fig. C.3

If λ is the resonant wavelength for a cavity of length l and radius a, we have for TE modes

$$1/\lambda^2 = 1/\lambda_c^2 + 1/\lambda_g^2$$

with

$$l = q\lambda_g/2$$

where $q = 0$, 1, 2 etc. and $\lambda_c = 2\pi/k_c$.

Since $k_c = p'_{mn}/a$ where p'_{mn} is the nth root of the equation $J'_m(k_c a) = 0$, we obtain

$$1/\lambda^2 = (k_c/2\pi)^2 + (q/2l)^2$$

or

$$1/\lambda^2 = \left(\frac{p'_{mn}}{2\pi a}\right)^2 + (q/2l)^2$$

If $D = 2a$ is the cavity diameter, then

$$1/\lambda^2 = (p'_{mn}/\pi D)^2 + (q/2l)^2$$

or

$$\lambda = \frac{1}{\sqrt{(q/2l)^2 + (p'_{mn}/\pi D)^2}} \quad \text{for TE modes}$$

and by similar reasoning p'_{mn} is replaced by p_{mn} for TM modes and we obtain

$$\lambda = \frac{1}{\sqrt{(q/2l)^2 + (p_{mn}/\pi D)^2}} \quad \text{for TM modes}$$

A very useful mode employed for wavemeters is the TE_{011} (H_{011}) shown in Fig. C.3.

The electric field lines are loops in the centre while the magnetic lines travel through them and down by the side walls. Hence, no currents flow between the end plate and walls. This is useful in a tunable cavity since movement of an end plate as a piston, does not disturb the field pattern. As there is a one half-wavelength variation of E along l, $q = 1$ with $p'_{mn} = p'_{01} = 3{\cdot}83$ and we obtain

$$\lambda = \frac{1}{\sqrt{(1/2l)^2 + \left(\dfrac{1}{0{\cdot}82D}\right)^2}}$$

as the resonant wavelength of the cavity.

Appendix D: Fibre mode theory[46,47]

The propagation of electromagnetic waves along an optical fibre can be analysed using Maxwell's equations for a dielectric medium, assuming the cylindrical coordinates r, ϕ, z where r is the radial direction, ϕ is the azimuthal direction and z is the direction of propagation along the optical fibre. The analysis which is somewhat similar to that for circular metal waveguides involves Maxwell's curl equations and the general wave equation for the electric field $\boldsymbol{E}$ and magnetic field $\boldsymbol{H}$. These are conveniently expressed by

$$\text{curl}\,\boldsymbol{E} = -\mathrm{j}\omega\mu\boldsymbol{H}$$

$$\text{curl}\,\boldsymbol{H} = \mathrm{j}\omega\varepsilon\boldsymbol{E}$$

$$\nabla^2\boldsymbol{E} = \gamma^2\boldsymbol{E}$$

$$\nabla^2\boldsymbol{H} = \gamma^2\boldsymbol{H}$$

where ω, μ, ε and γ have their usual meanings and ∇^2 is the Laplacian operator.

The application of these equations to an optical fibre using the appropriate boundary conditions, leads to the propagation of an infinite number of field patterns or modes which are either guided or unguided by the optical fibre. Only the guided modes which are mainly confined to the core material are of interest here. The analysis is simplified by assuming the 'weakly-guided' approximation $(n_1 - n_2) \ll 1$ where n_1 is the refractive index of the core material and n_2 is the refractive index of the cladding material.

Unlike metal waveguides which support pure TE or TM modes to satisfy the boundary conditions, a dielectric fibre supports only *hybrid* modes such as HE_{nm} or EH_{nm} modes depending on whether the H field is predominant or the E field is predominant respectively. The suffixes n, m refer to the number of field variations in the azimuthal and radial directions respectively.

Maxwell's equations for the 'weakly-guided' approximation lead to the wave equation

$$\frac{d^2\psi}{dr^2} + \frac{1}{r}\frac{d\psi}{dr} + \frac{1}{r^2}\frac{d^2\psi}{d\phi^2} + [n(r)k^2 - \beta^2]\psi = 0$$

where ψ refers to the E or H field parameter, $n(r)$ is the radial variation of the refractive index, $k = 2\pi/\lambda$ is the wave number and β is the propagation constant in the optical fibre.

The cylindrical geometry of the fibre leads to the general solution of the form

$$\psi = R(r)\cos(n\phi)\,e^{j(\omega t - \beta z)}$$

where the radial function $R(r)$ is obtained by solving the Bessel equation

$$\frac{d^2R(r)}{dr} + \frac{1}{r}\frac{dR(r)}{dr} + \left[(k_1^2 - \beta^2) - \frac{n^2}{r^2}\right]R(r) = 0$$

with $n = 0, 1, 2$ etc. and $k_1 = n_1 k$.

The required solution involves the well-known Bessel function of the first kind and of order n, if only guided modes are considered for which the field is finite at $r = 0$. Hence, we obtain

$$R(r) = J_n(ur) \qquad (r \leqslant a)$$

where
$$u = \sqrt{k_1^2 - \beta^2} = \sqrt{k^2 n_1^2 - \beta^2}$$

and
$$n_2 k < \beta < n_1 k$$

A propagating mode is cut off when it is no longer bound to the core and its field outside the core does not decay rapidly to zero. An important parameter connected with this cut off condition is the *normalised frequency* V defined by

$$V^2 = (u^2 + w^2)a^2 = \left(\frac{2\pi a}{\lambda}\right)^2 (n_1^2 - n_2^2)$$

where
$$u^2 = k_1^2 - \beta^2,\ w^2 = \beta^2 - k_2^2,\ k_2 = n_2 k$$

and a is the core radius.

The parameter V determines the number of modes a fibre can support. The number of modes that can propagate along the fibre may be conveniently expressed in terms of a normalised propagation constant b given by

$$b = \frac{a^2 w^2}{V^2} = \frac{(\beta/k)^2 - n_2^2}{n_1^2 - n_2^2}$$

Modes are therefore cut off when $w = 0$ or $\beta = n_2 k$ with the exception of the fundamental HE_{11} mode which has no cut off and ceases to exist only when the core radius a is zero. This is the basis on which the single-mode fibre is designed. Since the cut off condition for the next higher-order mode is given by $J_0(ua) = 0$, it yields a value of $ua = 2{\cdot}405$ from standard tables. Thus, we obtain

$$V = ua = 2{\cdot}405$$

and so single-mode operation is possible for $0 \leqslant V \leqslant 2{\cdot}4$. As the value of V increases, more modes can propagate along the fibre and for a multimode fibre, the number of modes N propagating is given by $N = V^2/2$.

Index

Admittance, 31, 35
APD, 100
Applegate diagram, 69
attenuation
  coefficient, 13
  optical fibre, 96
  waveguide, 63

Bends, 73
bolometer, 84
boundary conditions, 43
Brewster angle, 48
bunching, 69

Cavities
  cylindrical, 122
  rectangular, 121
cavity resonators, 119
characteristic impedance, 14
circular modes, 60
circulator, 119
coaxial cable, 2, 37
coefficient
  attenuation, 13
  phase-change, 13
  propagation, 13
  reflection, 27, 47
conducting medium, 39
critical angle, 48

Delay
  group, 24, 93
  phase, 24
detectors, 99
dielectric medium, 39
diode
  APD, 100
  edge-emitting, 88
  Gunn, 71
  laser, 88
  light-emitting, 11, 88
  PIN, 99
directional coupler, 75
directivity, 76
dispersion, 10, 92, 93, 95
distortion
  attenuation, 22
  phase, 22
distortionless line, 22

Electromagnetic
  field theory, 38
  waves, 37
ELED, 89

Ferrites, 116
fibre
  graded-index, 92
  step-index, 92
fibre mode theory, 123
field
  theory, 38
  vectors, 38
frequency measurements, 80

Group
  delay, 24
  velocity, 24, 52
Gunn effect, 71
gyromagnetic
  effect, 116
  ratio, 117

Hybrid
  ring, 74
  T-junctions, 73
hyperbolic solutions, 16

Impedance
  characteristic, 14

intrinsic, 42, 56
load, 33, 36
measurements, 81
normalised, 31
integrated optics, 105

Klystron
amplifier, 71
oscillator, 71
reflex, 69
two-cavity, 69

Lasers, 11, 89
LED, 11, 88
line
distortionless, 22
general, 12
high-frequency, 22
infinite, 15
loss-free, 20
low-frequency, 21
low-loss, 20
strip, 4, 115
two-wire, 1
loading, 23

Matched line, 17, 29
material dispersion, 92, 93, 95
Maxwell's equations, 38, 123
measurements
frequency, 80
impedance, 81
power, 84
VSWR, 81, 82
medium
conducting, 39
dielectric, 39
microstrip, 4, 115
modal dispersion, 92, 93
modes
cavity, 121
circular, 60
higher-order, 61, 92, 125
hybrid, 124
rectangular, 121
TE, 115
TM, 115

Normalised
admittance, 31
frequency, 124
impedance, 31
reactance, 30
resistance, 30
numerical aperture, 92

Optical
communications, 88
fibres, 10

Phase-change coefficient, 13, 45
phase
delay, 24
velocity, 24, 52
photodiode
avalanche, 100
PIN, 11
photodetectors, 99
PIN-FET receivers, 100
power measurements, 84
Poynting's theorem, 41
propagation coefficient, 13

Q-factor
loaded, 120
unloaded, 119
quantum efficiency, 100

Rayleigh scattering, 97
receivers, 99
rectangular modes, 121
reflection
coefficient, 27, 47
on lines, 26
refractive index, 47, 93
resonant wavelength, 122
responsivity, 100
rotary vane attenuator, 76

Sensitivity, 102
skin depth, 46
Smith chart, 30
standing wave detector, 77
stripline, 4, 115

TE waves, 8, 50, 115
TEM waves, 2, 4
test bench, 79
T-junctions, 73

TM waves, 8, 50, 115
two-wire line, 1, 37

VSWR, 29, 33, 80, 81, 82

Wave equation, 39
waveguide
  circular, 8, 60
  dispersion, 95
  rectangular, 8, 55
wavelength
  cut-off, 8, 54
  free space, 54
  guide, 54
  resonant, 122, 123
wavelength division multiplexing, 104
wavemeters, 77
waves
  electromagnetic, 37
  guided, 50
  reflected, 46
  refracted, 46
  standing, 27
  TE, 8, 50
  TEM, 2, 4
  TM, 8, 50
  travelling, 27